# Development of
# Biopharmaceutical
# Parenteral
# Dosage Forms

# DRUGS AND THE PHARMACEUTICAL SCIENCES

**Executive Editor**

## James Swarbrick

*AAI, Inc.*
*Wilmington, North Carolina*

### Advisory Board

Larry L. Augsburger
University of Maryland
Baltimore, Maryland

David E. Nichols
Purdue University
West Lafayette, Indiana

Douwe D. Breimer
Gorlaeus Laboratories
Leiden, The Netherlands

Stephen G. Schulman
University of Florida
Gainesville, Florida

Trevor M. Jones
The Association of the
British Pharmaceutical Industry
London, United Kingdom

Jerome P. Skelly
Copley Pharmaceutical, Inc.
Canton, Massachusetts

Hans E. Junginger
Leiden/Amsterdam Center
for Drug Research
Leiden, The Netherlands

Felix Theeuwes
Alza Corporation
Palo Alto, California

Vincent H. L. Lee
University of Southern California
Los Angeles, California

Geoffrey T. Tucker
University of Sheffield
Royal Hallamshire Hospital
Sheffield, United Kingdom

Peter G. Welling
Institut de Recherche Jouveinal
Fresnes, France

# DRUGS AND THE PHARMACEUTICAL SCIENCES

## A Series of Textbooks and Monographs

# Development of Biopharmaceutical Parenteral Dosage Forms

edited by

## John A. Bontempo

*Consultant*
*Biopharmaceutical Product Development*
*East Brunswick, New Jersey*

## CRC Press
Taylor & Francis Group
Boca Raton London New York

CRC Press is an imprint of the
Taylor & Francis Group, an **informa** business

CRC Press
Taylor & Francis Group
6000 Broken Sound Parkway NW , Suite 300
Boca Raton, FL 33487-2742

First issued in paperback 2019

ISBN-13: 978-0-8247-9981-6 (hbk)
ISBN-13: 978-0-367-40091-0 (pbk)

---

**Library of Congress Cataloging-in-Publication Data**

---

Development of biopharmaceutical parenteral dosage forms / John A. Bontempo, editor.
    p. ; cm. --(Drugs and the pharmaceutical sciences; v. 85)
  Includes index.
  ISBN-13: 978-0-8247-9981-6 (alk. paper)
  ISBN-10: 0-8247-9981-X (alk. paper)
  1. Pharmaceutical biotechnology.
  2. Parenteral solutions.
  I. Botempo, John A.  II. Series.

RS380.D48 1997
615'.6--dc21
                                   97-15956 CIP

---

Visit the Taylor & Francis Web site at
http://www.taylorandfrancis.com

and the CRC Press Web site at
http://www.crcpress.com

# Preface

Successful sterile product development of modern parenteral biopharmaceutical therapeutic agents demands the close interaction of interdisciplinary sciences encompassing molecular biology, fermentation, process development, protein chemistry, analytical biochemistry, pharmacology, toxicology, preformulation, formulation, clinical development, quality assurance, bulk manufacturing, packaging, sterile manufacturing, regulatory affairs, and marketing.

The physicochemical properties of proteins constitute the building blocks for preformulations and formulations development. Elucidation of these properties facilitates educated selection of compatible excipients to extend the shelf life of protein drugs.

The objective of formulation scientists is to make biopharmaceutical parenteral products that are safe, effective, pure, stable, suitable for production, cost effective, and marketable, and elegant. We cover liquid parenteral formulations of biopharmaceuticals. Lyophilized formulation development of similar biopharmaceuticals are more properly the focus and interest of other pharmaceutical scientists.

This book describes fundamentals and essential pathways for various formulation techniques, their purpose and function, and how each method is fundamentally related to and integrates with the other approaches to successful product development.

The book covers six key areas of the development of biopharmaceuticals for human use, and each chapter is written by one or more industrial scientists involved with the state-of-the-art techniques. Building on the key areas allows the formulator to incorporate specific tasks into biopharmaceutical design in order to reach the next plateau of development. The result is an understanding of the product development formulation process.

Following the introductory Chapter 1, Chapter 2 covers biopharmaceuticals currently licensed or in clinical development, including genetically engineered cells and engineered vectors. The fermentation process, the first key step in product development, is a key scientific and economic model. Once the goals of this process are defined, fermentation and harvest can begin. Quality must be designed into each step of the process to ensure success.

Chapter 3 discusses the purification and characterization techniques employed to produce a highly purified, economically focused, flexible process that can be transferred, scaled up for manufacturing, and validated to current regulatory standards. Several types of unit operations for isolation, purification, and characterization are also discussed. These are essential for the formulator to understand the biochemical structure of the active drug substance and to ensure quality of the final product.

The fourth and fifth chapters cover key phases of successful product development. Considerations are reviewed for drug delivery, formulation, stability studies programs, routes of deactivation and denaturation, aggregations, protein stabilizers, excipients, requirements of preservatives, and physicochemical properties of therapeutics. Attention is paid to how each is accomplished by the pharmaceutical scientists in order to enhance the success of the formulation.

The sixth chapter addresses basic concepts in analytical techniques, methods development, separation methods employing chromatographic and electrophoretic techniques, bioactivity methods covering bioassays, and immunoassays. The methods selected will show how to measure stability of biological activity.

Chapter 7 covers basic filtration theories, filter classifications and characteristics, filter performance criteria, validation and regulatory requirements, filtration systems, filter separation specific for recombinant protein and peptide processing, and future trends in filtration technology. This chapter will inform the formulator how to select the proper filters for each drug substance to maximize compatibility and minimize adsorption and inactivation.

The eighth chapter explains the physical, chemical, and toxicological properties of closures for parenteral products; protein adsorption on various elastomeric surfaces; strategies to reduce or eliminate adsorption; and specialized containers for biotechnological applications. The formulator is given a menu from which to select the most compatible elastomeric closure for a specific drug substance.

Our approach informs students, molecular biologists, protein chemists, process engineers, purification scientists, pharmaceutical formulators and manufacturing personnel, and those involved in QA/QC and regulatory affairs *how to perform* appropriate, systematic, sequential steps of formulation development.

## ACKNOWLEDGMENTS

I wish to express my sincerest appreciation to each of the contributors. I fully understand their overwhelming achievement in balancing the demands of their own work with fulfilling the additional responsibility that I asked of them for the successful completion of this book. My sincerest thanks to Tony, Paula, Basant, and Forrest for making this book possible.

I would also like to thank Dr. Edward J. Smith (Helvoet Pharma) for his constructive comments in reference to the chapter on elastomeric closures for parenteral biopharmaceutical drugs.

My special thanks also go to Drs. Patty Kiang (The West Co.), Y. John Wang (Scios Nova), and Pai Kao (Mayo Clinic), for giving me permission to include some of their work on protein adsorption on elastomeric surfaces. This addition made the book more complete.

Last but not least, my very personal gratitude and thanks to my wife, Loretta. Her skills in preparing my chapters were invaluable—considering the corrections, changes, deletions, and additions—as was her huge amount of patience and understanding. I dedicate this book to my friend, my wife. Thank you, Loretta.

*John A. Bontempo*

# Contents

# Contributors

**Forrest Badmington**   Senior Consulting Scientist, Millipore Corporation, Bedford, Masssachusetts

**John A. Bontempo, Ph.D.**   Consultant, Biopharmaceutical Product Development, East Brunswick, New Jersey

**Anthony S. Lubiniecki, Sc.D.**   Vice President and Director, Biopharmaceutical Development, SmithKline Beecham Pharmaceuticals, King of Prussia, Pennsylvania

**Paula J. Shadle, Ph.D.**   Director, Biopharmaceutical Quality Operations and Analytical Methods, SmithKline Beecham Pharmaceuticals, King of Prussia, Pennsylvania

**Basant G. Sharma, Ph.D.**   Senior Director, Bioanalytical Development, The R.W. Johnson Pharmaceutical Research Institute, Raritan, New Jersey

# 1

## Introduction to the Development of Biopharmaceutical Parenteral Dosage Forms

JOHN A. BONTEMPO
Biopharmaceutical Product Development, East Brunswick, New Jersey

## I. INTRODUCTION

Successful sterile product development of modern parenteral biopharmaceutical therapeutic agents demands close interactions of interdisciplinary sciences encompassing molecular biology, fermentation, process development, protein chemistry, analytical biochemistry, pharmacology, toxicology, preformulation, formulation, clinical development, quality assurance, bulk manufacturing, packaging, sterile manufacturing, regulatory affairs, marketing, and others.

The objectives of formulation scientists are to make biopharmaceutical parenteral products for human and veterinary use which are safe, effective, pure, stable, elegant, suitable for production, cost effective, and marketable.

Our goals in this book are to describe fundamentals and essential pathways for each scientific section, its purpose, function, and how each section is fundamentally related and how it integrates with the next section in the successful product development efforts.

This book covers six key areas with several chapters. Each of these are addressed by one or more industrial scientists involved with the "state-of-the-art" development of biopharmaceuticals for human use. Dividing the scientific contents into key areas allows the formulators to incorporate specific tasks into the biopharmaceutical design and allows them to reach the next plateau of development. When incorporated sequentially, this will result in the scientific understanding of the product development formulation process.

Chapter 2, Fermentation Process Events Affecting Biopharmaceutical Quality, covers in detail biopharmaceuticals currently licensed or in clinical development that include genetically engineered cell and engineered vectors. The fermentation process, the first key step in product development, is a key scientific and economic model. Once this target is defined, the fermentation and harvest system can begin. Quality is designed into each step of the process to ensure success.

Chapter 3, Development of Recovery Processes for Recombinant Proteins and Peptides, covers in detail the purification and characterization techniques and approaches to produce a highly purified, economically focused, flexible process that can be transferred, scaled up to manufacturing, and validated to current regulatory standards. Several types of unit operations, for isolation and purifications and characterization, are discussed which are essential for the formulator in the understanding of the biochemical structure of the active drug substance and to ensure quality of the final product.

Chapter 4, Preformulation Development of Parenteral Biopharmaceuticals, and Chapter 5, Formulation Development, cover key phases of successful product development. Considerations are reviewed for the drug delivery, formulation issues, stability studies programs, routes of deactivation and denaturation, aggregations, protein stabilizers, excipients, preservatives requirements, and physicochemical properties of therapeutics and how it is accomplished in order to enhance the formulation success by the pharmaceutical scientists.

Chapter 6, The Analytical Techniques, covers basic concepts in analytical techniques, methods development, separation methods employing chromatographic and electrophoretic techniques, bioactivity methods covering bioassays, and immunoassays. The methods selected will show how to measure stability of biological activity.

Chapter 7, Membrane Filtration Technology, covers basic filtration theories, filter classifications and characteristics, filter performance criteria, validation and regulatory requirements, filtration systems, filter separation specific for recombinant protein and peptide processing, and future trends in the technology. This section will inform the formulators how to select the proper filters specific for each drug substance to maximize compatibility and minimize adsorption and inactivation with the filter.

Chapter 8, Considerations for Elastomeric Closures for Parenteral Biopharmaceutical Drugs, covers the physical, chemical, and toxicological properties of closures for parenteral products, protein adsorption on various elastomeric surfaces, strategies to reduce and/or eliminate adsorption, and specialized containers for biotechnology application. The formulator is given a menu to select the most compatible elastomeric closure for the specific drug substance.

Our approach informs students, molecular biologists, protein chemists, process engineers, purification scientists, pharmaceutical formulators, manufacturing, QA/QC and Regulatory Affairs people *what to do*, and *how to do* appropriate systematic sequential phases of formulation development.

In this book we cover only liquid parenteral formulations of biopharmaceuticals. Lyophilized formulation development of similar biopharmaceuticals will be, we are sure, the focus and interest of other pharmaceutical scientists.

## II. KEY REQUIREMENTS TO CONSIDER BEFORE PREFORMULATIONS AND FORMULATIONS OF BIOPHARMACEUTICALS BEGIN

The applied formulation scientists today face formidable challenges in their quest to formulate stable recombinant protein therapeutics. Proteins possess unique characteristics. We are dealing with very complex, high-molecular-weight, highly purified, heat-unstable molecules, with a potential for aggregation. As this happens, chemical and physical changes occur; consequently, we can expect a great deal of instability, both physical and chemical. At this point, the formulation scientist must call on their knowledge of the various aspects of protein chemistry in order to design experimental approaches to prevent or control these processes. Protein instability mechanisms have been most recently reviewed by several investigators (1–10).

Chemical reactions such as oxidation, deamidation, proteolysis, racemization, isomerization, disulfide exchange, photolysis, and others will give rise to chemical instability. Physical instability will lead to denaturation, aggregation, precipitation, and adsorption.

It is of critical importance that when this happens, the denaturation mechanism(s) must be identified in order to select appropriate stabilizing excipients. These chemical excipients may be in the form of amino acids, proteins, surfactants, polyhydric alcohols, antioxidants, phospholipids, chelating agents, and others. Specific chemical and physical instability and potential stabilization approaches will be discussed in a later section.

## III. KEY PHASES FOR SUCCESSFUL INDUSTRIAL PRODUCT DEVELOPMENT OF BIOPHARMACEUTICALS

Successful industrial product development embodies several stages of experimental development. These phases represent pharmaceutical tasks that have to be clearly defined and performed for current Good Manufacturing Practices (cGMPs) and regulatory compliance. These pharmaceutical tasks are the focus of multidisciplinary and administrative groups. From my industrial experience, failure to put in place and work on these phases as a team can result in delays in accruing scientific data needed for a submission of specific regulatory compliance in a timely manner. These phases can be summarized as follows.

### A. Phase I

Demonstrate pharmacological activity of a crude fermentation protein broth and determine potential value.

## B. Phase II

Upon demonstration of pharmacological activity by a crude protein, appropriate data should be generated and compiled to support filing for an Investigational New Drug (IND).

## C. Phase III

The following experimental data should be obtained in order to determine if initial preclinical and clinical studies should be initiated: safety–toxicity, dose titration confirmation, and final safety studies.

## D. Phase IV

Upon evaluation of clinical data from phases I to III, a decision will be made to file a Product License Application (PLA).

## E. Phase V

During the Food and Drug Administration (FDA) review of the PLA, there will be questions by reviewers pertaining to specific issues. Depending on the nature of the questions, it will be the responsibility of that specific interdepartmental group to review and formulate answers for the FDA.

The key functions of each of these interdepartmental groups are described in Table 1 to clearly identify their respective responsibilities. Ultimately it will be the combined efforts of all of these interdepartmental groups that will impact on the acceptance or rejection of the product.

## IV. SELECTION OF DRUG DELIVERY SYSTEMS

Some of the key important factors in considering a specific delivery system are safety, stability, and efficacy. The parenteral administration of proteins and peptides today offers assured levels of bioavailability and the ability of the product to reach the marketplace first. It is safe to assume that over 98 percent of the protein therapeutics approved by the FDA today are injectable products, since parenteral administration avoids physical and enzymatic degradation.

However, intensive studies are in progress today in the industry for the application of alternate deliveries for proteins and peptides such as oral, nasal, transdermal, ocular, and oticular. In each case, membrane and gastrointestinal barriers must be extensively studied in order to increase bioavailability.

**TABLE 1**　Responsibilities of Interdepartmental Groups

|  | Responsibilities/tasks |
|---|---|
| 1. Molecular biology | • Expression in fermentation<br>• Yield improvement<br>• Alternate expression systems<br>• Plasmid characterization<br>• Support of regulatory requirements; postfiling, if necessary |
| 2. Fermentation development | • Initial fermentation<br>• Scale-up of crude material<br>• Production of required quantities<br>• Scale-up, yield, improvement, process validation<br>• Yield improvement |
| 3. Protein chemistry | • Initial characterization<br>• Continued characterization<br>• Support of regulatory requirements; postfiling, if necessary |
| 4. Process development | • Initial purification processes<br>• Production of required bulk active<br>• Scale-up<br>• Scale-up, yield, improvement, process validation |
| 5. Analytical development | • Development of appropriate bioassays<br>• Continuing assay development<br>• Perform required R&D assays<br>• Assay validation |
| 6. Analytical services | • Perform needed R&D assays<br>• Support stability studies<br>• Assay validation |
| 7. Pharmacology/toxicology | • Monitoring biological activity<br>• In vitro, in vivo screen for given indication<br>• Evaluation of dose titration, dose confirmation, completion of safety–toxicity studies |
| 8. Marketing | • Initial input into project<br>• Develop market plans<br>• Collaborate in the final product for launch and package design<br>• Final plans for launch |
| 9. Bulk manufacturing | • Production of needed fermentation/ purification material |

TABLE 1 Continued

| | Responsibilities/tasks |
|---|---|
| 9. Bulk manufacturing (continued) | • Physicochemical properties of active substance |
| 10. Preformulation/formulation | • Initial formulation development |
| | • Physicochemical properties of protein drug |
| | • Initial compatibility, screening, and stability |
| | • Selection of key excipients and initiate stability |
| 11. Sterile pharmaceutical manufacturing | • Sterile filling of needed material |
| | • Evaluate market package and product material |
| | • Scale-up, yield improvement and production of qualifying lots |
| | • Production launch, staff requirements, facilities for manufacturing of product for market |
| 12. Quality assurance | • Product specifications |
| | • Process validation |
| | • Material testing release |
| 13. Clinical development | • Conduct required studies |
| 14. Regulatory affairs | • All submission requirements to the FDA |
| 15. Front office | • It is of paramount importance to keep the front office routinely updated on the status of the product development. All lines of communication must remain open at all times. |
| | • There is no room for "surprises." |

There are several portals in considering the delivery of biopharmaceutical therapeutics, as shown below.

## A. Routes of Administration

1. Parenterally
   • Intravenous (I.V.)
   • Intramuscular (I.M.)

- Subcutaneous (S.C.)
- Others
2. Cavitational
   - Oral
   - Nasal
   - Ocular
   - Oticular
   - Vaginal
3. Respiratory
   - Nasal
   - Lung
4. Gastrointestinal
   - Oral
   - Rectal
5. Dermal
   - Topical
   - Transdermal

## B. Novel Technology for Controlled or Sustained Drug Delivery

Although the parenteral administration today represents the largest mode of protein drugs administration, alternate delivery systems of proteins and peptides are also being investigated with considerable efforts by a large number of biopharmaceutical companies. Some of these delivery systems are:

- Microencapsulation
- Implantable devices such as polymer matrix and bioerodable polymers
- Pumps for internal and external use

However, whatever delivery system may be selected, a great deal of focus should be on the *end-user*, either in a hospital or home environment. Alternate delivery systems, including the following, also pose unique problems associated with recombinant proteins and peptides.

1. High molecular weight
2. Not very lypophilic
3. Poor stability in solution
4. Usually small amounts available
5. Expensive
6. Adsorption problems

## REFERENCES

1. Hageman MJ. Drug Dev Indus Pharm 1988; 14:2047.
2. Wang Y-C, Hanson MA. J Parenteral Sci Technol 1988; 42:53.
3. Manning MC, Patel K, Borchardt RT. Pharm Res 1989; 6:903.
4. Geigert J. J Parent Tech 1989; 43:220.
5. Privalov PL, Gill SI. Adv Protein Chem 1988; 39:191.
6. Nozhaev VV, Berezin IV, Martinek K. CRC Crit Rev Biochem 1988; 23:235.
7. Pace CN, Shirley BA, Thompson JA. In: Creighton TE, ed. Protein Structure, A Practical Approach. Oxford: IRL Press, 1989:311.
8. Timasheff SN, Arakawa T. In: Creighton TE, ed. Protein Structure, A Practical Approach. Oxford: IRL Press, 1989:311.
9. Arakawa T, et al. Pharm Res 1991; 8:225.
10. Schein CH. Biotechnology 1990; 8:308.

# 2

# Fermentation Process Events Affecting Biopharmaceutical Quality

ANTHONY S. LUBINIECKI
SmithKline Beecham Pharmaceuticals, King of Prussia, Pennsylvania

## I. INTRODUCTION

Proteinaceous active ingredients of biopharmaceuticals are synthesized by biological expression systems and subsequently recovered, purified, and formulated. At each processing step the properties of the active ingredient can be modified and contaminants may be introduced. This chapter will attempt to describe some of the known phenomena that can occur during biological phases of processing; the following chapter will address those that occur during recovery and purification. The detection of these modifications to protein structure and their relevance to product quality will be discussed.

I use the term "biopharmaceuticals" to refer to purified proteins derived from recombinant DNA (rDNA) technology or to monoclonal antibodies (MAbs) derived from hybridoma technology. In the most fundamental sense, biopharmaceuticals have earned a place in modern medicine because of the exquisite specificity of their activity, which has proven impossible, so far, to reproduce in low-molecular-weight drugs. This specific biological activity reflects the three-dimensional structure of the protein molecule which in turn reflects at a minimum the primary amino acid sequence and posttranslational modifications (if any). The nature of the biological environment immediately after gene expression and translation can also be critical by providing the correct redox environment for disulfide formation, the appropriate scaffolding and glycosylation enzymes to permit proper posttranslational modifications, folding and/or transport, and so forth. In general, these factors must be considered at the time the biological system is constructed, and care must be taken to control the fermentation process to achieve consistent protein structure(s) and to maintain them during harvest and purification. Since the folded structure of proteins generally depends largely upon hydrogen bonding and solvent exclusion, which are weak forces averaging 5 to 10 kcal/mol, care must be taken throughout processing (fermentation, recovery, and formulation) to avoid physical or chemical treatments that lead to denaturation.

It is essential to understand those structural features of the protein of interest that are important to the desired activity and other properties of interest so that the correct biological system for expression can be chosen. Otherwise, it will prove difficult to attain and maintain the desired activity of the protein and product. This leads, inevitably, to the need for detailed characterization of the molecular structure of the protein of interest and the important structure-function relationships.

In general, small to medium-sized rDNA proteins are easily expressed intracellularly in microbial expression systems such as *Escherichia coli*. Key molecular features for successful expression in lower life forms includes relatively few disulfide bonds, relative resistance to proteases and lack of dependence on "humanlike" glycosylation for biological activity or pharmacological activity. In contrast, expression of secreted rDNA proteins or monoclonal antibodies in cells derived for higher eukaryotes generally do not face such limitations in terms of structure and size but proteolysis can be problematic even in cell culture systems. Examples of currently licensed/approved products and investigational products are shown in Tables 1 to 6. The worldwide market for these licensed/approved biopharmaceuticals in 1994 is estimated at about $7.5 billion or about 3 percent of the entire pharmaceutical industry.

This chapter will deal with biological considerations for products derived from rDNA and hybridoma technologies. Both the complexity of protein structure and the living processes of production lead to the requirement for detailed knowledge of the important properties of the protein in both functional and molecular terms in order to develop the proper manufacturing and quality strategy.

**TABLE 1**   Licensed/Approved rDNA Biologicals Expressed by Mammalian Cells

| Product | Protein | Cells | Year[a] | Licensed |
|---|---|---|---|---|
| Activase/Actilyse | tPA | CHO | 1987 | Broadly |
| Epogen/Procrit/Eprex | Epo | CHO | 1989 | Broadly |
| Epogen/Recormon | Epo | CHO | 1990 | Japan, Europe |
| Saizen | hGH | C127 | 1989 | Broadly |
| GenHevac B Pasteur | HBsAg | CHO | 1989 | France |
| HB Gamma | HBsAg | CHO | 1990 | Japan |
| Granocyte | G-CSF | CHO | 1991 | Japan, Europe |
| Recombinate | F VIII (80+90 kDa) | CHO | 1992 | Broadly |
| Kogenate | F VIII (80+90 kDa) | BHK-21 | 1993 | Broadly |
| Pulmozyme | DNase I | CHO | 1993 | Sweden, U.S., Switzerland |
| Cerezyme | Glucocerebrosidase | CHO | 1994 | U.S., Austria, New Zealand |
| Gonal-F | FSH | CHO | 1995 | Sweden, Finland |

[a]First licensure/approval.

TABLE 2    Animal Cell rDNA Proteins in Clinical Trial

| Protein | Clinical target |
|---|---|
| sCD4, sCD-41gG chimera | AIDS |
| gB, gD (HSV) | Herpes vaccine |
| gp 120, gp 160 (HIV) | HIV vaccine |
| tPA muteins | Thrombolysis |
| F VIII muteins | Hemophilia A |
| F VII | Hemophilia A with F VIII inhibitors |
| Chimeric or humanized MAbs against | |
|    her2 | Cancer |
|    CD4 | Autoimmunity, transplantation |
|    TNFα | Sepsis, rheumatoid arthritis |
|    CD20 | B cell lymphoma |
|    TAC | Transplantation |
|    CD18 | Hemorrhagic shock |
|    Leukointegrin | Multiple sclerosis |
|    CF54, CD7 | Rheumatoid arthritis |
|    RSV | Respiratory syncytial disease |
|    IgE | Allergy |
| Interferon beta (IFN-β) | Multiple sclerosis |
| Thyroid stimulating hormone (TSH) | Thyroid cancer |
| Soluble complement receptor-1 (sCR-1) | Acute respiratory distress syndrome |
| Transforming growth factor beta (TGF-β) | Soft tissue wounds |
| Soluble TNF receptor | Sepsis |
| Macrophage colony stimulating factor (M-CSF) | Cancer, hypercholesterolemia |
| Bone morphogenetic proteins (BMP-2, BMP-7) | Bone fractures |
| IL-6 | Thrombocytopenia, multiple sclerosis |
| Soluble IL-1 receptor | Allergy, asthma |
| IL-12 | Thrombocytopenia |

## II.  BIOLOGICAL PROCESSES

As brief review, proteins are translated by ribosomes from messenger RNA (mRNA) which have been transcribed from cellular genes (DNA). Usually, proteins are folded into three-dimensional structures; some types are exported while others are retained inside the cell. Modifications to protein

TABLE 3 Licensed/Approved MAb Biologicals

| Category | Product | Immunogen | Indication | Year | Licensed |
|---|---|---|---|---|---|
| Therapeutic | OKT3 | CD3 | GVHR | 1986 | Broadly |
| | Centoxin | Endotoxin | Sepsis | 1990 | Europe[a] |
| | ReoPro | Platelet IIb/IIIa | MI | 1994 | U.S., Europe |
| | Panorex | ? | Colorectal cancer | 1995 | Germany |
| In vivo diagnostic | Oncoscint O/V | CEA | Cancer | 1990 | Europe, U.S. |
| | Myoscint | Myosin | Cardiac | 1989 | Europe |
| Preparative | Roferon A | IFNα2A | Purification from cell lysate | 1986 | Broadly |
| | Monoclate | Factor VIII | Purification from plasma | 1987 | U.S. |
| | MonoNine | Factor VIII | Purification from plasma | 1992 | U.S. |
| | Kogenate | Factor VIII | Purification from conditioned medium | 1993 | U.S. |
| In vitro diagnostic | >100 | Various | Various | 1980 | Broadly |

[a]Withdrawn from marketing in 1993.

composition and structure may occur. This process is fundamentally similar in all cells studied from diverse phyla and kingdoms. However, some details vary among life forms that become important in heterologous protein expression. Frequently, the peculiarities of gene expression in a specific host organism interact with the properties of the protein of interest or its gene sequence. Some important ones that may impact product quality include codon usage in mRNA [relative to frequency distribution of transfer RNA

TABLE 4 Licensed/Approved Purified Natural Proteins

| Product | Protein | Cells | Year | License |
|---|---|---|---|---|
| Welferon/Sumitoferon | IFNα | Namalva | 1985 | Europe, Japan |
| Feron/Frone | IFNβ | HDF | 1980 | Japan, Germany |

TABLE 5   License Approved rDNA Biologicals—Microbial Expression Systems

| Product | Protein | Cells | Year | Licensed approved |
|---|---|---|---|---|
| Humulin | Insulin | E. coli | 1982 | Broadly |
| Protropin/Kabitropin | hGH | E. coli | 1985 | USA, Europe |
| Roferon A | IFNα2A | E. coli | 1986 | Broadly |
| Intron A | IFNα2A | E. coli | 1986 | Broadly |
| Recombivax/Engerix B | HbsAg | S. cerevisiae | 1986 | Broadly |
| Humatrope/Genotropin/ Nutropin/Eskatrope/ Norditropin | hGH | E. coli | 1987 | Broadly |
| Actimmune | IFNγ | E. coli | 1990 | USA |
| Neupogen | G-CSF | E. coli | 1991 | Broadly |
| Leukine/Prokine | GM-CSF | S. cerevisiae | 1991 | Broadly |
| Leukoprol | M-CSF | E. coli | 1991 | Japan |
| Proleukine | IL2 mutein | E. coli | 1992 | Broadly |
| Betaseron | IFNγ mutein | E. coli | 1993 | USA |
| Igef/Somazon | IGF1 | E. coli | 1994 | Sweden/Japan |
| Hanp | ANF | E. coli | 1995 | Japan |
| Humalog | Insulin mutein | E. coli | 1995 | Russia |

(tRNA) species for a given anticodon], signal leader sequence for secretion, protease profile and susceptibility, posttranslational covalent modifications (such as glycosylation, γ-carboxylation, acetylation, disulfide formation), folding, intracellular transport and secretion, as well as interaction with other proteins. Some of these topics are explored in further detail below.

## III. GENETIC CONSTRUCTION

For rDNA technology to be practical, the gene encoding the protein of interest is inserted into a vector allowing (1) vector replication in a suitable host (usually E. coli) prior to transfection into the ultimate host, (2) selection of the transfected host over non-transfected hosts, (3) in many cases, vector replication or amplification in the transfected host, (4) expression of the gene(s) of interest, and (5) intracellular transport of vector-specified proteins guided by signal sequences. A number of specific genetic elements

**TABLE 6** Investigational rDNA Biologicals—Microbial Expression Systems

| Product | Protein | Indication |
|---|---|---|
| PIXY 321 | CM-CSF/IL3 chimera | Thrombocytopenia and neutropenia |
| | SOD | Oxygen toxicity in premature infants |
| | EGF | Eye surgery |
| | FGF | Wound healing |
| | IGF1 | Osteoporosis, diabetes |
| | NGF | Peripheral neuropathy |
| | PDGF | Wound healing |
| | Consensus IFNa | Hepatitis C |
| | IL11 | Thrombocytopenia |
| | IL1a | Prevention of bone marrow suppression |
| | IL3 | Bone marrow transplantation |
| | IL4 | Immunodeficiency, cancer |
| Alvircopt Sudotox | sCD4/PE40 chimera | AIDS |
| | TNF | Cancer |
| Antril | IL-1 RA | Rheumatoid arthritis |
| Auriculin | ANF | Acute kidney failure |
| Glucagen | Glucagon | Hypoglycemia |
| | BPI-23 | Sepsis |
| | Hemoglobin | Blood substitute |
| | SLPI | Cystic fibrosis |
| | SCF | Adjuvant chemotherapy |
| | HSA | Shock |

are required to accomplish these various tasks; many must be tailored to specific hosts for optimal results. For example, promoters to facilitate transcription of the gene of interest are usually strong, closed, and inducible for expression in *E. coli* (such as pL or lacZ) but are constitutively open (and usually of viral origin) for expression in mammalian cells. Aminoglycoside antibiotic resistance genes are commonly used for selecting transformed host cells but different antibiotics are employed depending upon the host chosen (typically, kanamycin or neomycin for *E. coli* and G418 for animal cells). Occasionally, quality control of the purified biopharmaceutical is enhanced by careful choices of vector components. In several cases, co-purification of other proteins encoded by the expression vectors could be detected in the purified biopharmaceutical and elimination of the contaminant from the product depended upon reengineering the vector to remove the offending gene. Thus, solutions to quality problems with biopharmaceuticals often lie a long way from the point where the problem is detected.

In addition to choosing the correct genetic elements for a given task, it is essential to employ well-characterized gene sequences of known source, order, composition, and function to construct vectors. These pieces of DNA can be viewed as a type of master raw material where quality can be directly controlled only once—before transformation of the host. In addition to documentation of source and function of the vector elements, it is important to confirm the arrangement of the pieces by restriction mapping. The location and function of every open reading frame or coding region should be known. Failure to utilize well-known genetic elements may lead to the expression of gene sequences of unknown function as well as the possibility of regulatory delays while the true structure and function of the uncharacterized DNA sequences is elucidated.

Finally, the nucleic acid sequence of the DNA region encoding the protein of interest is determined for the plasmid to be used for transfection as well as for the transfected host cell (at the master cell bank stage) and typically once more for an aliquot of postproduction cells to confirm the identity of the same construct. In some cases the development and validation of a quantitative peptide map may substitute for sequencing of postproduction cells.

It may be important to balance the utilization of degenerative codons in mRNA to the tRNA prevalence of the host chosen for expression of heterologous proteins. For example, the codon preferences for amino acids like leucine and arginine are different between *E. coli* and vertebrates. Failure to consider the relative abundance of various tRNAs may lead to a pause in translation (waiting for a rare charged tRNA species to appear) which may in turn lead to frameshift or to premature termination. Misincorporation may also occur at frequencies as high as 32% (Lys for Asp in MS2 protein expressed in *E. coli* in nonstringent strains) under starvation conditions or as high as 12% (Lys for Arg in IGF1 expressed in *E. coli*) when codon usage is not optimized to prevailing tRNA abundance (1,2). Molecular size determinations of purified proteins by SDS-PAGE or by chromatography is often able to detect premature termination as well as initiation at an internal methionine. C-terminal and N-terminal sequence analysis respectively is used to confirm the location of the affected site(s). Peptide mapping is an excellent tool to detect frameshifts and misincorporation, and to assist in locating sites of premature termination or internal initiation.

## IV. GENE EXPRESSION

There is a concern that incorrect amino acids may be incorporated into the protein of interest due to mutation, mistranscription, or mistranslation of the transfected gene, thereby giving rise to product-related molecules which might be toxic or immunogenic. This subject was extensively discussed at two recent symposia (3,4). Briefly, the probability that at least one mutation will occur in at least one copy of the transfected gene during fermentation is quite high. However, the mutant protein is unlikely to be a significant portion of the expressed protein unless this mutation also confers a major selective advantage on the mutant in terms of relative growth rate during brief fermentations, or unless a mutation conferring a minor growth rate advantage occurs during transfection or banking followed by hundreds of cell divisions required to prepare the protein of interest (5).

It is also true that mutation is a fundamental biological process; mutation and natural selection is widely accepted as the pathway by which fish evolved into humans. Further, mutation occurs every day in the human body. It may be possible to promulgate regulations outlawing mutation but it is not possible to achieve compliance by Mother Nature. Fortunately, life is well adapted to survive under circumstances where proteins show evidence of modest levels of mutation, mistranscription, and mistranslation. For example, it has been estimated that the average human is heterozygous for 3 to 5 genetic traits which, if homozygous, would result in death or in visible organic disease (6). In other words, all of the cells of each human express 3 to 5 mutated genes such that half of the resultant encoded proteins are incorrect, usually without obvious problems. Thus, mutation is not only natural, it is also not necessarily harmful.

There are also data to suggest that mistranslation occurs in *E. coli* at rates of about $2 \times 10^{-4}$ to $2 \times 10^{-3}$ (7). Overexpression in *E. coli* of selected genes may lead to an increased frequency of mistranslation. Overexpression of murine EGF, which contains no Phe codons in *E. coli*, led to EGF containing 1.1% Phe residues (8). Workers at Monsanto showed that translation errors increase 10-fold in lacZ expression under conditions typically used for induction in *E. coli* compared to uninduced conditions (9). There is generally limited data for animal cells on such concerns but there is also little rationale for expecting major differences from *E. coli*.

While mutations, mistranscriptions, and mistranslations unquestionably occur and lead to errors in primary sequences, barriers exist to prevent their appearance in product vials. Many proteins with changes in primary structure fail to fold properly. Misfolded proteins are generally not secreted but degraded by cytoplasmic proteosomes. Purification procedures with multiple orthogonal modes of separation are another barrier to contamination of final product by aberrantly expressed proteins. Builder has commented that the multi-column purification procedure for tissue plasminogen activator (tPA) failed to purify a tPA mutein containing only three different amino acids out of 527 (6). Rohde reported that the purification process for G-CSF removes three minor contaminant molecules in which mistranslations occur due to third base "wobble" in His codons resulting in incorporation of Gln at positions 53, 157, or 171 (10). While each of these contaminants constituted less than 0.5% of G-CSF extracted from inclusion bodies, they were not detectable in final product.

It is believed that the studies normally performed to characterize the expression system and protein, to validate the process, and to confirm the quality of each product lot are adequate to assure protein integrity from concerns regarding aberrant expression. Among the most powerful tools to detect aberrant product-related molecules with single amino acid changes is peptide mapping which can detect such molecules when they achieve a relative abundance of 4 to 8% (11) or less under special circumstances. By contrast, gene sequencing cannot detect point mutations until they reach a relative abundance of 15% (12) and cannot detect this transcription or mistranslation at any level. To date no licensed product prepared from a validated process has manifested evidence of misexpression at the product level. A single case exists wherein an investigational protein (from a non-validated process) was observed by standard peptide mapping analysis to contain low but detectable levels of an aberrant protein sequence which was eventually traced to a point mutation (13). Currently, there is no evidence to link any known adverse reactions or immune responses to biopharmaceuticals to expression problems. In summary, while there is evidence that expression errors exist and theoretical concern that expression errors might harm product recipients, current data suggest that available methods are adequate to control the problem to acceptable levels.


## V. CELL PHYSIOLOGY

Cell physiology during cultivation should be optimized to provide the maximum amount of recoverable, qualitatively acceptable protein. This, of course, assumes that the unacceptable impurities in the harvest can be removed and that the remainder are innocuous. Much has been written describing quantitative relationships between optimization of cell physiology and harvested protein to which the curious reader is referred (14–16).

However, I would like to concentrate here on some qualitative effects of cell physiology on the protein of interest.

Sublethal starvation of cells for essential nutrients may have a variety of subtle metabolic and biological effects. One of the best examples is methionine deprivation in *E. coli* which induces the expression of an enzymatic pathway for norleucine synthesis. Norleucine is also capable of being charged by some Met-tRNA species and, thus, of being misincorporated for methionine. In one study of bovine somatotropin (bST) expression in *E. coli*, 14% of genetically specified methionine residues in bST were randomly substituted by norleucine (17). The activity of several enzymes are unaffected by replacement of methionine by norleucine (18–20), although this might only mean that Met residues are not involved in the active site or otherwise do not influence enzymatic activity. In practice the problem is easily prevented by providing adequate methionine levels throughout the fermentation.

A major category of physiological events that impacts product quality is posttranslational processing events. They include proteolysis (removal of signal sequences, processing of proenzymes and prohormones like proinsulin), sulfation (Factor VIII), phosphorylation (DNase), gamma carboxylation (Factors VII and IX, Protein C), N-terminal sequence blockage by cyclization of glutamate to pyroglutamate (many antibody heavy chains), glycosylation, and many others. These modifications are accomplished by enzymes which are under genetic control of the host. If the activity or pharmacology of a given protein requires a specific posttranslational modification, it is essential that the host cell chosen for expression be capable of providing the appropriate enzymes to accomplish the task. For example, plants, yeast, and insects are all capable of glycosylation but tend to add mostly high mannose forms rather than the complex forms typical of mammalian cells. This is important for therapeutic proteins where pharmacokinetics are important, since mannose-endcapped proteins are frequently subject to rapid clearance by asialo-GM1 receptors present on Kupfer cells (fixed macrophages of the liver). Another potential mechanism for clearance from the circulation is natural or acquired antibody directed against the glycosylated moieties. Murine cells tend to add galactose subunits in a peculiar conformation (21) which reacts with natural antibodies present in most humans to the extent of consisting of 1 percent of circulating immunoglobulin (22). The existence of these "natural" antibodies in man is attributed to the presence of the same carbohydrate structure in *E. coli*, a major human intestinal microbe. Thus, murine cells may not be ideal hosts for the preparation of glycosylated biopharmaceuticals intended to possess long plasma half-life. However, it is probably not important for an in vivo imaging MAb or FAb where the desired activity occurs in minutes following administration.

Another class of posttranslational processing is proteolysis. All proteins have genetically determined leader sequences which are cleaved after translation and transport. Factor VIII is translated as a 330,000 Da protein which is eventually processed to yield two smaller molecules of 80,000 and 90,000 Da, respectively, which are found in the product vial. The original manufacturing process for recombinant insulin utilized two separate *E. coli* fermentations, one for A chain and one for B chain. Each peptide was isolated and purified separately, mixed together and then, chemically oxidized to form mature insulin because *E. coli* has a strongly reducing internal environment. Later processes express the A and B chains as a chimeric fusion protein with an internal spacer region which can be cleaved later. A summary of posttranslational modifications to some marketed rDNA products is shown in Table 7. Thus, expression hosts must be chosen with posttranslational processing in mind.

Posttranslational modification may interact with cell physiology; for example, the glycosylation pattern of expressed proteins may change when mammalian cells are deprived of glucose (23). In addition to maintaining proper concentration of the building blocks of glycosylation, it is also im-

**TABLE 7**  Glycosylation of Animal Cell Products

| Protein | N-linked | O-linked | Comment |
|---|---|---|---|
| tPA | Complex, high mannose | Yes | High mannose site essential to PK |
| EPO | Complex | Yes | Required for PK |
| hGH | No | No | Unusual for secreted protein |
| G-CSF | No | Yes | Required for resistance to aggregation and heat denaturation |
| FVIII | Complex | | |
| DNase | Complex, high mannose | No | Mannose is phosphorylated |
| Cerebrosidase | Complex | No | Targeted to macrophages by conversion of complex to high mannose forms |
| FSH | Complex | No | Required for in vivo activity |

portant to provide a consistent environment for the cells in an effort to have reproducible glycosylation patterns. It is also known that different fermentation systems (such as hollow fiber reactors) are more prone to provide reduced glycosylation (24). This probably reflects the nonhomogeneous conditions of hollow fiber reactors in which the cells at the distal end of the apparatus are exposed to medium substantially depleted of nutrients and oxygen by cells at the proximal end of the device (where fresh medium first enters the system).

Further, glycosylation patterns may change during the course of batch suspension culture fermentations. For example, the addition of complex multiantennary N-linked carbohydrates to ASN[97] in IFNγ expressed by CHO cells is reduced after logarithmic cell growth stops but that at ASN[25] is not (25). The nature of product of the glycosylation process will reflect the relative abundance and activity of the many enzymes involved as well as the concentration of the key reactants. Excessive glutamine consumption by transamination may give rise to excess $NH_3$ which may drive intracellular pH upward, outside the optimum range for sialyl transferase, resulting in reduced sialylation of GM-CSF by CHO cells (26). Altered glycosylation may affect the potency or safety of glycoproteins in a variety of ways—by altering properties such as biological activity, tissue distribution, pharmacokinetics, antigenicity, or immunogenicity.

The microheterogeneity resulting from glycosylation may also frustrate analytical attempts to detect important events such as instability on storage. The chief tools for routine quality control of carbohydrate in glycoproteins is compositional analysis of the hydrolyzed sugars and ionic exchange chromatography. For research characterization purposes and nonroutine troubleshooting, mass spectrometry and capillary zone electrophoresis is gaining favor. Tools for other types of posttranslational modification include isoelectric focusing, N-terminal protein sequencing, and specific chemical assays as appropriate for other changes. Using these techniques many laboratories report that validated fermentation processes are capable of providing glycoproteins with consistent glycosylation from batch to batch (5,27–29).

## VI. MEDIUM COMPONENTS

Some medium components can directly interact with the protein of interest. One major category of such interactions is enzymatic action. Enzymes of cellular origin from medium components may interact with the active ingredients. For example, proteases liberated from dead or dying host cells may cleave the peptide backbone of protein of interest or lead to N- or C-terminal fraying (or into the cell lysate process fluid for intracellular proteins expressed in microbial hosts). Similarly, glycosidases liberated by CHO cells such as sialidase may remove sialic acid residues from rDNA-derived proteins (30). DNase obtained from late stages of batch fermentations of CHO cells showed evidence of reduced sialic acid content due to enzymatic removal of sialic acid as well as reduced phosphate due to enzymatic removal of phosphate as well as evidence of deamidation (31). Depending on the importance of the carbohydrate to the in vivo action of a given protein product, this may have no effect (cytokines), some effect (tPA) or major effect (erythropoietin). It is the responsibility of the process development team to evaluate the existence and importance of such phenomena. Such reactions are frequently important in the determination of when to harvest the fermentor (before protein of interest is substantially modified) or how to conduct a unit operation like storing cell paste (quickly freezing and to very low temperatures).

Another set of interactions is covalent modification. If tPA is expressed in the presence of bovine serum, two things happen: (1) some of the molecules are cleaved enzymatically at position 276, and (2) some of the molecules are covalently bound to naturally-occurring bovine serum protease inhibitors (32). The existence of such hybrid molecules with many of the features of tPA (but with the antigenicity of bovine proteins) could significantly complicate downstream purification and affect product safety; therefore, animal serum used to propagate cells was removed prior to initiation of tPA collection. Low-molecular-weight medium components can also affect product quality. tPA has 35 cysteine residues. One of these is unpaired and capable of forming disulfides with free thiols such as cysteine when present at significant concentration (>3–6 mM). This can be observed upon isoelectric focusing analysis of purified tPA. Thus, in such cases when the protein of interest has one or more unpurified thiols, medium modification to improve the biological aspects of fermentation (i.e., increasing the cysteine content of medium) may change the qualitative nature of the product (31).

In addition to covalent modifications, rDNA-derived proteins also interact noncovalently with other classes of chemical substances. Many proteins possess metal ion binding sites. Even 316 L stainless steel used ubiquitously for preparation of biopharmaceuticals corrodes slowly, releasing low levels of constituent metal ions. Once bound, proteins may continue to carry these ions despite downstream processing, including extensive diafiltration. Some types of protein bind lipids, such as hepatitis B surface antigen and albumin. The lipid "contaminant" profile of these proteins following purification usually reflects the lipid profile of the recombinant host organism. For vaccines and proteins administered at low dose this phenomenon has no likely significance. For proteins administered at high dose any concerns can be evaluated as part of the preclinical safety assessment in animals.

It is useful to consider whether the protein of interest may interact with other components of the cell or the medium (or buffer) based upon the known properties of the protein of interest. In addition to looking for enzymatic and covalent interactions, non-covalent interactions should also be considered, not only with soluble components but with equipment surfaces as well. Some hydrophobic proteins like IFNβ bind to some plastic surfaces while other proteins may bind to glass or metal surfaces or to filter membranes. Such considerations may be especially important in hydrophobic proteins lacking secondary structure, where noncovalent interactions may lead to denaturation. Physical forces may interact with and may alter protein structure by causing aggregation or denaturation by hydrodynamic shear (e.g., by bubble action in sparged fermentation) or excessive heat (during the mechanical lysis of rDNA-modified microbes to extract protein). If such interactions are not considered in advance and prevented or minimized by intelligent process design, altered molecules may contaminate the final product. Typical analytical procedures which detect those types of problems are SDS-PAGE, isoelectric focusing, and various types of HPLC.

## VII. BIOLOGICAL CONTAMINANTS

The most medically significant contaminants of the previous generations of biological products have been biological contaminants—microbes and their constituents. Initially, bacteria and fungi (and their toxic byproducts) were the major contaminants of plasma products and vaccines but as aseptic processing methods and detection systems improved, this group of contaminants became relatively rare. Subsequently, the principal contaminants of conventional biological products of human or animal origin were viruses; that pattern remains true to this day. This subject is extensively reviewed elsewhere (33,34) but a few brief comments will be made here.

Viral contaminants of plasma products (and organ donations and transfusions) have killed thousands of recipients in the United States. Before the current generation of diagnostic tests facilitated safety assessment of the blood and organ supply before use by identifying contaminated donors, tens of thousands of recipients developed hepatitis every year. Since 1980, approximately 10,000 hemophiliacs have become infected with HIV. During World War II, nearly 30,000 cases of hepatitis were associated with a series of lots of yellow fever vaccine stabilized with unpasteurized human albumin given to military personnel.

Thus, the history of previous generations of biological products shows ample evidence of the effects of inadequate controls on raw materials and on process conditions. Better diagnostic technology has substantially reduced these risks; in addition for most (but not all) products, purification steps and inactivation procedures can be validated to reduce specific safety concerns. The new generation of microbially expressed rDNA products does not use raw materials of human origin and animal viruses do not replicate in bacterial and yeast cells; therefore, the risk of viral infection in recipients of such products prepared under current Good Manufacturing Practices (cGMP) is virtually nil.

The new generation of cell culture–derived biopharmaceuticals clearly uses animal cells which are capable of supporting replication of animal viruses, including some human pathogens and most use raw materials of animal origin. Safety is assured by four separate mechanisms. These include the use of characterized cell banks and certified raw materials, the use of process validation to remove or inactivate potential contaminants, the use of appropriate assays to screen for viral contaminants, and procedural and engineering controls on the manufacturing process, equipment, and facilities. As reviewed elsewhere (35), these methods have proven extremely effective. To date no evidence of viral infection exists attributable to contamination of any rDNA or monoclonal antibody product from cell culture, despite their use in approximately one million product recipients.

Two related considerations are currently subjects of regulatory activity. The first is the use of ingredients or components of human origin in biopharmaceuticals. For example, erythropoietin, OKT-3 MAb, and IFNα are formulated in human albumin. Human transferrin has been employed in some serum-free media for animal cell culture. Validation studies can be performed to demonstrate that specific risk factors such as HIV would not survive processing. However, additional new European Union (EU) regulations seem to require that all ingredients of human origin be identified on product labelling along with statements of compliance with current plasma donation standards, a summary of validation data, and a statement regarding the advisability of specific vaccinations such as hepatitis B. Given the speed with which mandated plasma screening assays are changed due to technological improvements, the lag between making and using a plasma-derived ingredient to prepare a biopharmaceutical with a significant shelf life, and the continued emergence of new viral pathogens, this regulatory initiative may encourage manufacturers to replace all human source ingredients as rapidly as possible.

Another concern is bovine spongiform encephalopathy (BSE), a progressive degenerative neurological disease of cattle related to scrapie of sheep and similar diseases of man and other mammals. It has reached epidemic proportions in the UK where over 100,000 cattle have been destroyed from 1988 to 1994. It is likely to be spread by the practice of using cattle and sheep protein (offals) as an oral dietary supplement without proper treatment (heat and organic solvent). These same conditions of animal husbandry exist in many other countries, including the United States. BSE is not known to be transmissable to man but it can be transmitted by parenteral route to chimpanzees, ostriches and cats. This prompted CBER in 1991 to request biological manufacturers to report the use of bovine and ovine raw materials in licensed or investigational products, including the national origin of the raw material and procedures used to minimize the risk of BSE and scrapie. Of 601 product responses, 63 percent were reported to utilize ovine or bovine ingredients (36). In 1993 a few weeks after detecting BSE in Canada in a cow imported from the UK, FDA promulgated regulations for drugs and biologicals which restrict use of bovine-derived raw materials to those from BSE-free countries (37). EU has additional regulations regarding bovine ingredients; recently, Germany has promulgated new regulations which describe conditions for accepting products with ingredients of bovine, ovine or caprine origin. These may eventually force biopharmaceutical manufacturers to find alternative sources for animal-derived raw materials such as insulin, transferrin, meat hydrolysates, and enzymes, among others.

## VIII. PHARMACOLOGICALLY ACTIVE CONTAMINANTS

A concern since the dawn of modern biotechnology has been the possibility of immunological reaction to host cell protein contaminants. Humans have natural antibodies to *E. coli*, yeast, and mammalian cells like CHO. Historically, host or medium components have been implicated in allergic reactions to vaccines. Serum sickness is a known sequela of heterologous plasma protein administration. Administration of highly purified biopharmaceuticals to humans has not resulted in these types of allergic reactions associated with conventional biologicals; observed clinical reactions seem nonspecific (minor pain/redness at injection site, cytokine release syndrome).

In general, the level of host cell proteins in biopharmaceuticals are very low ($<1\%$) and there is no direct evidence that they cause problems in product recipients. There is evidence indicating that the incidence of non-neutralizing antibody to the administered protein may be higher when less highly purified preparations are administered to man (38). However, it is not clear whether the responsible immunogen is product-related contaminants or host cell proteins.

Some rDNA-derived human proteins are immunogenic but some are not (39). Generally, these antibodies have no clinical significance for most human rDNA products (except IFNα, where relapse of hairy cell leukemia correlated with appearance of neutralizing antibody) (40). While about 25% of hemophilia A patients rDNA-derived Factor VIII develop neutralizing antibodies, multiple independent mutations are known to give rise to hemophilia A. Thus, any given Factor VIII gene product will be a "foreign" protein to some fraction of hemophilia A patients. Increasing the dose of Factor VIII in such patients seems to provide therapeutic effect (41). Mab products of murine origin induce human antimouse antibody (HAMA) responses in recipients which did become therapeutically limiting but the several human MAbs evaluated in early clinical studies do not have this problem. Chimeric Mabs still seem to generate HAMA responses to their murine variable regions. The rDNA ("humanized") Mabs (possessing only murine CDRs) are just beginning clinical trials; immunogenicity for many of these is predicted to be acceptable. Immunoassays and SDS-PAGE are used to monitor host cell protein contaminants on a lot-to-lot basis. The immunogenicity of human proteins cannot be evaluated meaningfully in animal preclinical studies; therefore, a major element of clinical trials for biopharmaceuticals is the assessment of immune response to the administered protein in recipients, usually by an immunoassay especially configured for this purpose.

No evidence exists of other proteins from the host cell producing pharmacological effects in product recipients. While some biopharmaceuticals manifest toxicity in recipients, these affects are due to the inherent pharmacology of the active ingredient(s), not the result of pharmacological action by unrelated contaminants or degradation products of the active ingredient. (The degradation of proteins give rise to amino acids which are nutrients. Degradation of posttranslational modifications gives rise to other nutrients such as sugars or to other types of biochemicals normally found in humans.) This stands in sharp contrast to chemical drugs where contaminants and degradation products often are responsible for observed toxicity.

## IX. SUMMARY

Biopharmaceuticals depend upon living biological systems to create complex macromolecules with defined three-dimensional structures for optimal activity. Assurance of product quality depends in part upon choice of the proper host cell to correctly synthesize and process the desired proteins, upon fermentation process development to maintain protein integrity after synthesis, and upon process validation to assure protein consistency from lot-to-lot. Fermentation processes are designed and operated to prevent biological contamination of the protein and to optimize product safety. Examples of biological phenomena which can affect final product quality are given and appropriate solutions and control procedures are discussed.

## REFERENCES

1. Parker J, Johnson TC, Borgia PT. Mol Gen Genet 1980; 180:275.
2. Seetharum R, Heeren RA, Wong EY, et al. Biochem Biophys Res Commun 1988; 155:518.
3. Various authors Biologicals 1993; 21:89–156.
4. Genetic stability and recombinant product consistency. In: Brown F, Lubiniecki AS, eds. Develop Biol Standard. Basel: Karger, 1994:83.
5. Wiebe ME, Lin NS. In: Lubiniecki AS, Vargo SA, eds. Regulatory practice for biopharmaceutical production. New York: Wiley-Liss, 1994:33.
6. Horaud F, Lubiniecki AS. Robertson J. Develop Biol Standard 1994; 83:187.
7. Rosenberger RF. Develop Biol Standard 1994; 83:21.
8. Scores CA, Carrier M, Rosenberger RF. Nucleic Acid Res 1991; 19:3511.
9. Bogosian G, Violand NB, Jung PE. Kane JF. In: Hill WE, Dahlberg A, Garrett RA, Moore PB, Schlesinger D, Warner JR, eds. The ribosome. Washington DC: American Society for Microbiology, 1990:546.
10. Rohde MF, Lu HS, Rush RS. Develop Biol Standard 1994; 83:121.
11. Burstyn D, Copmann T, Dinowitz M, et al. Biopharm 1991; 4:22.
12. Garnick R, Helder JC. Develop Biol Standard 1994; 83:177.
13. O'Connor JV, Keck RG, Harris RJ, Field MJ. Develop Biol Standard 1994; 83:165.
14. Lubiniecki AS. In: Lubiniecki AS, Vargo SA, eds. Regulatory Practice for Biopharmaceutical Production. New York: Wiley-Liss, 1994:165.
15. Thomas J. In: Lubiniecki AS, ed. Large Scale Mammalian Cell Culture Technology. New York: Marcel Dekker, 1990:93.
16. Arathoon R, Hughes T. In: Chin YY, Gueriguian JL, eds. Drug Biotechnology Regulation. New York: Marcel Dekker, 1991:177.
17. Bogosian G, Violand BN, Dorward-King EJ, Workman W, Jung PE, Kane JF. J Biol Chem 1989; 264:531.
18. Anfinsen CB, Corley LB. J Biol Chem 1969; 244:5149.

19. Nadir F, Bohak Z, Yariv J. Biochemistry 1972; 11:3202.
20. Giles AM, Marliere P, Rose T, et al. J Biol Chem 1988; 263:8204.
21. Galili U, et al. Proc Natl Acad Sci USA 1987; 84:1369.
22. Borrebaeck CAK. Immunol Today 1993; 14:477.
23. Szkudlinski MW, Thotakura NR, Bucci I, et al. Endocrinology 1993; 133:1490.
24. Hayter PM, Curling EM, Baines NJ, et al. Biotechnol Bioeng 1992; 39:327.
25. Hooker AD, Goldman MH, Markham NH, James DC, Bull AT, Jenkins N. Society for General Microbiology Symposium, Bath, UK, 1995.
26. Anderson DC, Goochee CF, Cooper G, et al. Glycobiology 1994.
27. Lubiniecki AS, Anumulla K, Callaway J, et al. Develop Biol Standard 1992; 76: 105.
28. Adamson SR, Charlebois TS. Develop Biol Standard 1994; 83:31.
29. Boedeker B. Society for General Microbiology Symposium, Bath, UK, 1995.
30. Warne TG, Ching J, Ferreri. J, et al. Glycobiology 1993; 3:455.
31. Sliwkowski MB, Gunson JV, Cox ET. Advances in Pharmaceutical Bioproduction. New Orleans, 1992.
32. Lubiniecki AS, Arathoon R, Polastri G, et al. In: Spier RE, Griffith JB, Stephenne J, Crooy P, eds. Advances in Animal Cell Biology and Technology for Bioprocesses. London: Butterworth, 1989:442.
33. Harris GS. The Hazards of Immunization. London: Aithone Press, 1967:324.
34. Horaud F, Brown F, eds. Develop Biol Standard 1991; 75.
35. Lubiniecki AS, Wiebe ME, Builder SE. In: Lubiniecki AS, ed. Large Scale Mammalian Cell Culture Technology. New York: Marcel Dekker, 1990:515.
36. Albrecht P. Develop Biol Standard.
37. Food and Drug Administration. Letter to Drug Manufacturers, December 17, 1993.
38. Fleming GA, Jordan AW, Chin YY. In: Chin YY, Guerguian TL, eds. Drug Biotechnology Regulations. New York: Marcel Dekker, 1991:341.
39. Zoon K. Symposium on Genetic Stability and Recombinant Product Consistency, Annecy, France, 1993.
40. Von Wussow, P. Symposium on Genetic Stability and Recombinant Product Consistency, Annecy, France, 1993.
41. Gomperts ED, Courtea SE, Lynes MD, et al. Develop Biol Standard 1995; 83: 111.

# 3

# Development of Recovery Processes for Recombinant Proteins and Peptides

PAULA J. SHADLE
SmithKline Beecham Pharmaceuticals, King of Prussia, Pennsylvania

## I. INTRODUCTION

The purification of recombinant proteins and peptides which are to be used
as parenteral products aims to produce a highly purified and biologically
active product in a consistent manner. The goals of process development
and validation are to develop a reproducible process that accomplishes the
purification goals at reasonable yield and cost, and can be validated to current regulatory standards. This reproducible process should also assure a
high degree of confidence that the final product will meet specifications for
purity, potency, stability, and other quality attributes (Points to consider,
1994; Points to consider, 1993; CPMP Guidelines; Federici, 1994). This
chapter will provide an overview of the approaches that are used to develop

and characterize an isolation and purification process, and will also discuss the types of information collected that are most useful to the formulation scientist. In addition to the in-process analytical characterization of the product described here, the reader is urged to refer to Chapter 6, on final product characterization.

The end product of the cell culture or fermentation process is the feedstream that enters the isolation and purification process. This feedstream contains the recombinant product along with impurities consisting of product-related degradation products, fragments, or aggregates, and contaminants such as host cell proteins, DNA, or adventitious agents. Additional impurities may be added in the form of process chemicals during fermentation and purification. It is the goal of the purification process to separate the desired protein from both impurities and contaminants and obtain a purified product in an active conformation which is not immunogenic or highly toxic.

The operating conditions of the fermentation, changes in cell metabolism, or genetic instability of the vector may all affect the product and the efficacy of the purification process. Degradation of the product may begin at the time it is synthesized and continue throughout isolation and purification. Cellular errors in metabolism that have been reported in the production of recombinant proteins include the substitution of norleucine for methionine in bacterial fermentations (Lu et al., 1988; Bogosian et al., 1989), variant glycosylation (Maiorella et al., 1993; Parekh et al., 1987, 1989a, b), and mistranslation errors that are sequence dependent. Many reports have indicated that glycosylation varies with the cell line, media composition, time in culture, and between perfusion and stirred tank culture fermentation systems (Prior et al., 1989; Lai and Strickland, 1987; Parekh et al., 1987; Robinson et al., 1993). Those variants of a product whose generation cannot be prevented by the fermentation scientist or molecular biologist enter the purification process, and may need to be separated from the intact product at considerable yield loss.

During isolation and purification, the protein product is exposed to a variety of physical and chemical conditions which may also catalyze its degradation. Process chemicals which are used may be either deleterious to the product's potency or stability, or harmful to the patient if present in the final product. Understanding the purification process aids the formulation scientist in developing stability-indicating methods, anticipating analytical difficulties, and determining what contaminants and impurities should be specifically evaluated in the final product.

## II.  ISOLATION METHODS

Isolation is the process of separating the protein product from cells and cell debris at the end of the fermentation period (reviewed in Marston, 1986 and Uhlen et al., 1992). The product may be present in the media or within the cells, and it may be in a soluble or an insoluble form, depending on its primary structure, the nature of the expression system, and sometimes the extent of expression. Isolation is technically challenging because the feed-stream is a complex mixture of DNA, lipid, protein, and media components, the volume is large, and the product may be insoluble or unstable in the feedstream. Scaling up the process may be challenging because small scale equipment may be different in design from large scale equipment, and because scaling up the cell culture process may result in significant changes to the feedstream (van Reis et al., 1991).

Changing the host cell line or the expression vector may affect the purification process by altering both host cell-derived contaminants and product-related degraded protein. For example, the product may move from the soluble to the insoluble fraction; a new protease may be expressed which degrades the product. The purification challenges are also related to the chemistry and stability of the product itself, making previous experience with the parental cell line not fully predictive of future processes. Because of the admixture of proteases, media components, and other impurities, a diverse array of degradation reactions can occur during isolation (see Table 1). Degradation of the product may have already begun during fermentation, or it may be unique to particular steps in the recovery process. Degradation may be increased when a new holding step is placed between

**TABLE 1**   Types of Product Degradation Seen During Isolation Steps

| Degradation | Examples | Reference |
| --- | --- | --- |
| Fragmentation, chemical or proteolytic | Tissue plasminogen activator (t-PA) | van Reis et al., 1991; Builder et al., 1988 |
| Oxidation | Interleukin-2, basic fibroblast growth factor | Kunitani et al., 1986a, b; Thompson and Fiddes, 1991 |
| Deamidation | Human growth hormone, triose phosphate isomerase | Reviewed in Gracy, 1992 |
| Aggregation | Interferon-gamma | Reviewed in Wetzel, 1992 |
| Precipitation | | Reviewed in Wetzel, 1992 |

fermentation and purification, for example by storing cells at –70°C or holding conditioned media in a tank prior to purification. A large initial volume and low product concentration can exacerbate the problem by increasing the contact time of the product with harmful components in the feed before and during concentration.

Slight variations in the operation of the fermentation can result in changes to the composition of the feed, and affect the behavior of the product or contaminants in isolation steps. Since degradation products, once formed, may not be easily removed during purification, the effectiveness of the isolation procedure may set the limits on the homogeneity and stability of the final product.

## A. Isolation Strategy for a Soluble Product

When the product itself is soluble and secreted by the host, the isolation method needs to process large volumes and capture the product from a relatively dilute feedstream, while removing proteases or other agents that could damage the product (Figure 1). Typically, this is done by using

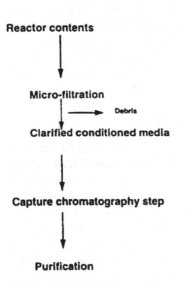

**Reactor contents**

**Micro-filtration** → Debris

**Clarified conditioned media**

**Capture chromatography step**

**Purification**

FIGURE 1  Flow diagram of a typical isolation scheme used when the product is soluble and secreted into the culture medium. The purpose is to rapidly isolate the product in a reduced volume and remove any degrading substances from product contact.

microfiltration or centrifugation to remove debris and whole cells, followed by chromatography that "captures" product via ion exchange, affinity, or some other strong interaction (Prior et al., 1989; van Reis et al., 1991; Builder et al., 1988). The primary goal is to remove any contaminants such as proteases which could degrade the product. At the same time, volume reduction is performed to concentrate the product in a much smaller volume suitable to downstream processing. The clarification step may combine tangential flow and depth filtration methods or may involve centrifugation, depending on the amount and type of debris present. The method is optimized for yield and time of operation. The product's stability in the conditioned media should be studied to qualify suitable holding and processing times.

> **Example 1.   Tissue plasminogen activator.**   Tissue plasminogen activator (tPA) was expressed recombinantly in CHO cells and purified from the conditioned media (van Reis et al., 1991; Builder et al., 1988). This recombinant protease was found to be degraded by a protease present in the fermentation broth which converted tPA to a two-chain form which was not fully active and was difficult to remove by purification. The process was modified to inhibit the enzymatic activity of the protease and to remove other inhibitory factors from the media. The early diafiltration and chromatography steps were designed to separate the contaminating protease from tPA (Builder et al., 1988). These process modifications reduced the concentration of two-chain degraded product to a minimal level.

## B.  Isolation Strategy for an Insoluble Product

Some products are expressed in an insoluble form, such as in bacterial inclusion bodies (Marston, 1986; Ohlen et al., 1992; Kane and Hartley, 1988). To purify these products, a multistage isolation strategy is common (Figure 2). First, intact cells are separated from the culture media in a volume reduction step, usually by microfiltration or centrifugation. Lysis of the cells is done by one of several methods to make product accessible to solubilizing agents (Kane and Hartley, 1988; Goeddel et al., 1979; Szoka et al., 1986; Tsuji et al., 1987). Often, the product can be substantially purified and separated from host cell proteins and proteases while it is protected by its insolubility (Abour et al., 1994; Martin, 1991; Ogez et al., 1980; Schoner et al., 1985; Williams et al., 1982). A series of different extraction conditions may be useful to remove DNA, lipid, host cell proteins, and degraded or fragmented product. Detergents, chaotropes, and various salt and pH

insoluble product

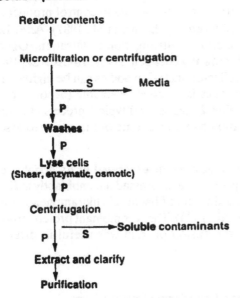

**FIGURE 2** Flow diagram of a typical isolation scheme used when the product is intracellular and insoluble. Volume is reduced with cells maintained intact, if possible. The precipitate containing the product may be washed with buffers to remove media, then cells are lysed to release product. Soluble contaminants may be further removed before the product is solubilized and clarified. The solubilized product is now ready for further purification. Y, yes; N, no.

conditions have been used on different inclusion body preparations. Developing these steps is empirical, since some products will partition into the extracted and the insoluble fraction, resulting in variable product loss.

Mechanical shear, osmotic shock, and enzymatic digestion with lysozyme have all been used successfully to induce cell lysis (Marston, 1986; Uhlen et al., 1992; Goeddel et al., 1979; Szoka et al., 1986). Mechanical methods generate heat and shear, which may denature, precipitate, or partially solubilize some proteins. Viscosity can be a major problem in subsequent steps since nucleic acid is released from lysed cells. The nucleic acid may vary greatly in size as the result of variable shear during isolation, as well as with scale, treatment, and the plasmid copy number of the cells. Enzymatic treatments have been useful in this regard (Tsuji et al., 1987;

Abour et al., 1994; Martin, 1991). In choosing a suitable method, the consistency of cell lysis across scale-up and the ability to control product stability during lysis must be considered (Keshavarz et al., 1987). After lysis, the inclusion bodies are collected by centrifugation or filtration (Ogez et al., 1989; Keshavarz et al., 1987; Mannweiler et al., 1989; Hoess et al., 1988).

Selective extraction of collected inclusion bodies can be highly effective for purification, if the product is relatively insoluble (Schoner et al., 1985; Benson et al., 1993). Table 2 lists several typical process chemicals that may be used during isolation steps, and some of the concerns associated with them.

**Example 2.   Human lung surfactant expressed in** *Escherichia coli.* Human lung surfactant protein was expressed recombinantly in *E. coli* as a fusion protein with a domain of bacterial chloramphenicol acetyltransferase (Benson et al., 1993). The recombinant protein product was hydrophobic, and the fusion protein was useful in directing

**TABLE 2**   Typical Process Chemicals Used During Isolation Steps

| Chemical | Purpose | Concern to product quality |
|---|---|---|
| Urea | Solubilization, inhibit aggregation | Carbamylation of product |
| Guanidine HCl | Solubilization, inhibit aggregation | Presence in final product Denaturation of product |
| Dithiothreitol, beta-mercaptoethanol | Prevent or reverse oxidation | Presence in final product as a mixed disulfide |
| Chelating agents (EDTA) | Inhibit proteolysis and metal-catalyzed oxidation | |
| Detergents | Solubilization, extraction, inhibit aggregation | Trace contamination in final product; peroxides in Tween may oxidize product |
| Enzymes: lysozyme, benzon nuclease | Lysis, nucleic acid digestion (reduces viscosity) | Selectivity of enzyme Enzyme contamination of product |
| Salts: ammonium sulfate, PEG | | May introduce metal contamination and oxidize product Residual in final product |
| pH changes | May activate enzymes | Proteases, carbohydrases Denaturation of product |

expression into inclusion bodies. The inclusion bodies, containing fusion protein, were isolated in the pellet fraction after cell lysis, and extensively purified by selective extractions with low concentration of guanidine hydrochloride and detergents. The enriched pellet containing the fusion protein was then solubilized using chaotropic conditions.

Yield losses during preparation of inclusion bodies are usually due to nonspecific losses during handling, and to a partitioning of the product into soluble and insoluble fractions. A protein product may partition when irreversible degradation or denaturation occurs, or when extraction methods do not allow adequate time to fully solubilize the product. Many proteins are protected throughout isolation by the use of antioxidants and reductants. Operating conditions such as low temperature and rapid processing times also are helpful to inhibit irreversible oxidative damage, misfolding, aggregation, and other types of degradation (Szoka et al., 1986; Tsuji et al., 1985; Abour et al., 1994; Martin, 1991; Ogez et al., 1989; Schoner et al., 1985; Williams et al., 1982; Keshavarz et al., 1987).

**Example 3. Recombinant human interleukin-2 (rIL-2) expressed in *E. coli*.** During purification and refolding of rIL-2 expressed in *E. coli*, a minor peak appeared which was shifted in retention on RP-HPLC (Kunitani et al., 1986a). Modified rIL-2 in which a single Met codon was replaced by an Ala codon was homogeneous on RP-HLPC (Kunitani et al., 1986b). The minor species was identified as Met sulfoxide, apparently generated by oxidation side reactions that occurred during isolation, purification, and refolding. Buffer additives and controlling other process conditions minimized the generation of this oxidative degradation product.

A soluble intracellular product may require lysis followed by clarification to remove debris before the product stream is ready for chromatography. Typically, some form of tangential flow microfiltration is used, and depth filtration may also be required. Alternatively, selective precipitation with ammonium sulfate or polyethylene glycol can be very effective in inhibiting proteolysis while concentrating product (Gilligan et al., 1987; Kamihira et al., 1992; Scopes, 1982). Again, the use of these process chemicals might add to analytical needs to test for trace residuals in final product. In practice, precipitation has not been used often in biotechnology products in manufacturing, but the practice is very common during research phase and early scale up.

Recent technology developments in isolation and purification include the use of fluidized or expanded bed chromatography columns, in which the chromatography resin is expanded rather than being packed (Erickson et al., 1994; Chase and Draeger, 1992). The expansion allows debris and buffer to pass through in a plug flow, rather like a piston, without clogging or pressurizing the column, while product binds selectively to the resin. Fluidized bed chromatography may allow a single column operation to replace a filtration step plus a chromatography column. Vendors are just beginning to offer columns designed for this purpose, which requires more careful process control and may result in a longer contact time of product with degrading enzymes. Another method that is being explored is to prepare filters having covalently linked ion exchange or affinity groups on them that can perform clarification and chromatography in one step (Goffe et al., 1988). Direct batch adsorption to ion exchange resin during solubilization was also performed by Hoess et al., 1988.

The main disadvantage which all of these approaches share is that column capacity may be greatly reduced compared to a packed column bed. Batch adsorption, being an equilibrium method, also is less efficient, lower in yield, and lower in resolution. However, the ability to function in the presence of debris is significant and may mitigate all of these disadvantages.

## C. Optimization of Isolation

Optimization of isolation steps usually includes three major steps:

1. Minimizing the contact time of product with harmful contaminants and process chemicals
2. Controlling temperature, pH, and other parameters to inhibit degradation of product
3. Adding substances that inhibit chemical degradation and denaturation, such as reductants, chelating agents, enzyme inhibitors, detergents, and chaotropes

The isolation and capture steps also are optimized to accept variation in the volume, concentration, and purity of the feedstream that comes out of fermentation, and to reduce that variability so that a consistent feedstream reaches the downstream process. The target purity and homogeneity of the final product must be carefully defined in order to set the purification goals.

## D. Isolated Product

When the isolation stage is complete, the product is preferably free of debris and live cells and concentrated in a buffer suitable for performing chromatography. In many cases, product stability is not assured at this stage, and the protein can only be held for a limited time and often only at low temperature. Degradation of the product during isolation translates into yield losses or increased product heterogeneity in the purified bulk product, both undesirable consequences. This explains the use of numerous process chemicals to prevent oxidation (dithiothreitol and EDTA for example), to protect product by chemical modification (sulfonation and citriconylation), to inhibit proteolysis (chelating agents, and other inhibitors), and to reduce or inhibit aggregation (chaotropes, specific buffers, and other process parameters). The safety concerns of these process chemicals and the probability that residuals are present in the final product should be considered when devising plans for final product release testing, validation, and formulation development.

## III. NONCHROMATOGRAPHIC METHODS OF PURIFICATION

To achieve a high degree of purification of proteins and peptides, chromatography has been the method of choice for recombinant proteins because of its scalability, high resolving power, and efficiency. However, chromatography is capital-intensive, requires skilled labor, and is an empirical science. As column chromatography is scaled up to meet demands of larger markets for high-dose products, equipment limitations and the cost of precision pumping and mixing systems become noticeable and may suggest that other methods be evaluated. Nonchromatographic methods, which are the primary purification methods used in the chemical, oil, and food industries, are more likely to be employed for clarification, volume reduction, and buffer exchange in recombinant protein purification. Recent examples of the use of nonchromatographic methods to achieve purification suggest that these methods do have a place in biotechnology.

## A. Filtration

Although commonly used to remove particulates and to reduce volume or exchange buffers, filtration as a purification method has not been frequently employed. A cascade of ultrafiltration (UF) membranes of varying molecular weights could be used to fractionate a feedstream by molecular weight, as was reported by van Reis and Builder, 1992. By recirculating the feedstream over the first UF membrane and passing the permeate through a second membrane having a smaller molecular weight cut-off, the system achieved removal of molecules both smaller than and larger than the product simultaneously, with minimal monitoring and intervention required. Such a system would be well adapted to a continuous processing mode of operation and potentially can achieve a high purification factor because of the iterative nature of the purification. Stability of some products might be affected by the longer operating time, during which the product continues to be exposed to the contaminants, and to shear.

Future developments of new filter chemistries, including affinity membranes, may make fractionation by filtration of significant value in purification process development and manufacturing. To date they have been used most effectively at small scale (Nachman et al., 1992), while manufacturing processes have largely relied on the high selectivity and reproducibility of chromatography for achieving high standards of purity.

## B. Selective Precipitation

Selective precipitation can be an effective mode of purification. Ethanol and other agents have been used to precipitate clotting factors from blood (Schwinn et al., 1989; Hrinda et al., 1989), and various agents have been used to prepare and wash bacterial inclusion bodies (see Section II.B). Precipitation reduces the volume and may enhance the purity of the product. Ammonium sulfate precipitation has the added advantage of inhibiting the activities of several proteolytic enzymes, thus rendering product more stable. Organic solvent precipitation has been used to delipidate tissue extracts and to inactivate viruses in blood-derived products (Hrinda et al., 1989).

Polyethylene glycol can also be used to precipitate proteins selectively by salting-out (Scopes, 1982, Chapter 2), and also may promote protein refolding (Cleland et al., 1992). Isoelectric precipitation was used to selectively precipitate an IgM at its pI, and achieved a high degree of purity at large scale (Steindl et al., 1987). The precipitate was collected by centrifugation and then resolubilized.

Both recombinant and nonrecombinant human insulins can be selectively precipitated in a crystallization reaction (Steiner & Clark, 1968). The addition of zinc ion to soluble, refolded human insulin expressed in *E. coli* induced formation of a hexameric complex consisting of six insulin molecules with zinc ion in the center. Misfolded species or aggregated protein were excluded from the crystallization reaction, and the precipitate was highly stable (Chance et al., 1981).

Selective precipitation presents some interesting contrasts to chromatography. Properly optimized, precipitation has achieved the high selectivity of chromatography, as the insulin example demonstrates. However, some precipitation unit operations use organic solvents (for example, ethanol) and generate toxic wastes that are costly to dispose of. Collecting a precipitate may require an expensive continuous centrifuge when done at manufacturing scale, thus may not offer capital or labor advantages over chromatography. The selectivity of the method may not be adequate to prevent contaminants coprecipitating with product. Alternatively, product may partition during precipitation and result in yield reduction and variation in both yield and purity. The presence of contaminants and impurities in the feed stream may change the precipitation properties of the product, making batch to batch variation an issue (Wetzel, 1992). Since these effects may be related to the scale and to the quality of the feed, they are difficult to predict from lab bench scale experiments. In many processes, careful adjustment of the feed concentration and composition have been the keys to ensuring success and reproducibility of a precipitation step.

Two-phase extraction is a related technique, in which a product may be separated from contaminants by its ability to partition selectively when a phase separation is induced in the solvent. Two-phase extraction may be less subject to nonspecific product losses than a precipitation method, because all components remain soluble during the separation. In a recent report, recombinant insulin-like growth factor 1 (IGF-1) was purified to 97% purity at high yield by adding polyethylene glycol and sodium sulfate to inclusion bodies that had been solubilized with chaotrope at high pH. The purification operation could be performed in the harvest tank and the solution phase containing purified protein was collected with simple equipment, clearly an advantage compared to column chromatography (Builder, 1994).

When precipitation is used in a process, it is important to evaluate the effects on product stability and homogeneity. Table 3 summarizes some types of product degradation that may be encountered, especially during filtration and precipitation steps of some proteins, and the factors that influence these types of degradation. Process optimization seeks to minimize these effects and set safe limits for operation of the process.

## IV. CHROMATOGRAPHIC PURIFICATION METHODS

The purification of a recombinant protein is usually accomplished through a series of chromatographic operations, which separates the product from

**TABLE 3** Degradation Occurring During Filtration and Precipitation Steps

| Type of degradation | Issues/causes |
|---|---|
| A. Seen in filtration steps | |
| Oxidation of sensitive residues: (Met, Cys-SH and S-S bonds; also His, Trp, Lys, others) | May result from aeration of sample |
| Aggregation and precipitation | Catalyzed by shear, aeration, heating, and high concentration |
| Proteolysis and chemical fragmentation | Affected by contact time, temperature, buffer conditions, and feed composition |
| B. Seen in precipitation steps | |
| Incomplete precipitation | Variable yield or purity, caused by feed variation, product heterogeneity |
| Product degradation | If product co-precipitates with a protease or other degradant; affected by time, temperature |
| Product denaturation | Caused by unfolding in the buffer, high local concentration in the precipitate, shear during collection of precipitate, time |
| Process chemicals | Additives become contaminants, and may be difficult to remove from the product |
| C. Seen in two phase extractions | |
| Process chemicals | Additives may be difficult to remove from the product; may degrade product (carbamylation) |
| Precipitation at interface | May inhibit separation, reducing purity and yield; can affect scalability of process |

contaminants and impurities. Contaminants include host cell proteins, DNA, and lipids, media components, and other process chemicals. In addition, impurities such as altered product that is inactive, reduced in potency, immunogenic, or otherwise toxic should be separated from intact product. In practice, it may be very difficult to determine the potencies and toxicities of the many product variants which are inherent to most proteins. Furthermore, the formulation scientist needs to receive a relatively homogeneous product in order to be able to evaluate its stability and identify its degradation products. Therefore, the purification scientist customarily seeks to minimize the heterogeneity of the impurity profile.

The general strategy in devising a purification process is to exploit the physical properties of the intact product in order to separate it from contaminants and impurities (Sitrin et al., 1986; Jungbauer, 1993). Figure 3 summarizes the general stages of purification process development. The purification goals are defined operationally, a series of separation steps are surveyed for feasibility and then the promising ones are linked. The resulting process is then optimized by incremental changes that improve yield, purity, and practicality. Last, process validation verifies that the process can be operated in a state of statistical control, and can purify product which is consistent in its yield, purity, potency, and stability. Because process

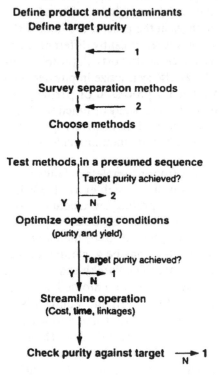

FIGURE 3   Flow diagram of the steps typically taken in developing a purification process for a recombinant protein. When target goals are not met, the scientist iterates to an earlier step in the flow diagram. Thus, many process changes may occur during development to optimize first purity, then scalability and economics of the process.

optimization and development are often concurrent with assay development, new information generated by improving assays of product purity may trigger redevelopment of the process steps at any time.

## A. Surveying a Purification Unit Operation

The goal of the first stage of developing a unit operation is to identify unit operations that selectively remove the key contaminants and impurities (Jungbauer, 1993). The type of separation to be tested is related to the biochemical characteristics of both product and contaminants. If product and contaminants differ greatly in charge, ion exchange chromatography is a natural choice; if in hydrophobicity, then hydrophobic interaction chromatography (HIC) or reversed phase liquid chromatography (RP-LC) may be useful, and so on. Purification of the same recombinant protein from a new expression system may require changes in the process methods if the composition of contaminants or impurities is significantly different. Contaminants may include such entities as nucleic acids, host cell proteins, lipids, and putative viruses. The host cell may also synthesize impurities such as product which has been posttranslationally processed or partially degraded, and product homologues, which are genetically or functionally related to the product. Homologues are of concern if they can be expected to copurify with the product and be significantly more immunogenic or have biological activity.

In addition, product variants resulting from incomplete or faulty processing during synthesis (see Chapter 2, as well as Hsieh et al., 1983; Lubiniecki, 1990; Parekh et al., 1987, 1989a, b), proteolysis, oxidation, and other post-translational events may need to be removed. Because product variants are very similar in chemistry to the product itself, their removal may be more difficult. Variants in glycosylation may be generated during cell culture (Maiorella et al., 1993; Hsieh et al., 1983; Parekh et al., 1989a). Products that are refolded must be purified to remove unfolded and misfolded variants (Marston, 1986; Uhlen et al., 1992; Seely and Young, 1991). Product degradation can occur both during fermentation and during isolation and purification, and degradants may need to be purified out (Mariani and Tarditi, 1992; Federici, 1994). For example, protein aggregates are usually carefully controlled because they may be immunogenic or have different pharmacokinetics in vivo (reviewed in Wetzel, 1992).

Finally, some process chemicals that are added during fermentation or purification are potentially toxic impurities and must be removed during purification (Quinlan et al., 1992). For example, pharmacologically active components added to cell culture media, ligands leached from affinity columns, or redox and chaotropic agents may need to be removed in a downstream unit operation either for reasons of safety or because of deleterious effects on the stability of the product (Quinlan et al., 1992; Levine et al., 1992; Federici, 1994). The potential presence of such impurities should be considered as early as possible during the development of a process, since the preferred order of chromatography steps may be affected. For example, an affinity column which leaches a toxic ligand is a poor candidate for the last step in a process, regardless of how well it performs in other ways.

During scouting, a separation is performed using assumed conditions with a small number of variations and the outcome in purity and yield is measured with any available assay. Promising methods are then chosen for further optimization. Multiple parameters, such as pH, ionic strength, and flow rate, often affect the outcome and synergize. The choice of chromatography support may affect separations significantly and in ways that are not fully predictable. The process scientist works closely with analytical methods scientists to devise assays which can detect impurities, contaminants, and product degradation, and a quantitative in-process assay for product.

## B. Optimizing a Separation

Optimization is best done by systematically varying key parameters to define the operating conditions which result in acceptable yield, purity, potency, and stability. Dynamic capacity is defined by performing breakthrough analyses at various flow rates (Jungbauer, 1993). Other important parameters may include column length, buffer composition, and buffer counterions, and temperature. Each of these parameters may affect the purity of the eluate, which may indicate changes in the stability and/or potency of the in-process intermediate or the final product. For example, high purity may decrease stability if the mechanism of degradation is self-association (aggregation). If the product degrades primarily via proteolysis at a given intermediate step, high purity would increase its stability. Each product and each host cell is different.

The goals of process development are to maximize yield at acceptable purity, while identifying the sensitive parameters. Once this is done, it is very important to look for opportunities to simplify and streamline the operation of each step (Figure 3). Reductions in the number of buffers used to run and regenerate a column, or even in the volumes used, may translate to cost savings when the process is scaled up (Seely and Young,

1991). Improvement of column cleaning conditions may mean higher product purity and process consistency. However, these changes could affect product purity, potency, or stability.

With the availability of chromatography supports that can support very high flowrates, process scientists now have the potential to cycle a small column multiple times per batch of product if desired. This strategy may be appropriate when the process must be scaled up in existing equipment, when chromatography resin is not available in large quantities, or when a very expensive affinity resin is used, since a small column can process large amounts of product per hour. Careful calculations are needed to determine the optimal size and optimal flow rate of the column. Flow rate and dynamic capacity will interact with column size to determine the time required to perform the operation, the number of cycles per batch. and the expense of cleaning. This approach will usually increase QC and validation costs because the process scientist must prove that such multiple cycles generate product which is equivalent in purity, yield, potency, and stability (Reisman, 1988; Sofer and Nystrom, 1989).

A unit operation is considered optimized when general operating ranges for the critical parameters have been found within which yield and purity are acceptable. The true optimum may not be the best place to operate a process if purity or yield drop off steeply around this point (Seely and Young, 1991). Operating at such an optimum might mean that small fluctuations in controlling the operation, such as are likely to occur during production, could affect the final product yield, purity, potency, or stability. Process settings at which small fluctuations in process parameters do not significantly affect the quality and quantity of product produced make the process more "rugged." The steepness of the decay in process outputs such as yield or purity can be used to measure and define process ruggedness (see Figure 4).

## C. Types of Chromatographic Separations

Several types of chromatographic separations are available for the purification of proteins. The main types of separations discussed here are effective, can be scaled up to process and manufacturing scales, and have been employed in several protein products. Their advantages and disadvantages are discussed below and summarized in Table 4 (Jungbauer, 1993). The use of orthogonal methods is usually the most efficient means to achieve maximal purification in the smallest number of unit operations. Since some types of operations may expose the protein to extreme conditions, sensitive products may degrade during purification. This degradation is either avoided by

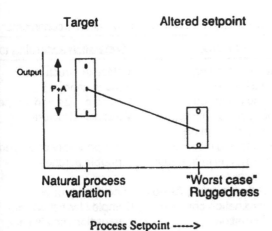

Process Setpoint ----->

**FIGURE 4** Diagram illustrates the concepts of process variation and process ruggedness. When the process is performed "exactly" at the setpoint conditions of pH, flow rates, and other critical operating parameters, the yield, purity, potency, and stability will still vary somewhat. This variation, called "natural process variation," represents the uncontrollable fluctuations in parameters due to equipment design, calibration, and so on. When the setpoint of the process is offset from the usual setpoint but kept within the permitted range, the impact of this altered setpoint upon yield, purity, potency, or stability is a measure of process ruggedness. The larger the decay (a steeper line above) in the outputs measured, the less rugged the process is, and the more important it is to control the input pH, etc., to its intended setting. P + A, process (P) plus assay (A) variation contributions sum up to create the measured process variation.

substituting a different separation method, or is inhibited, purified out, or shown to not affect potency and pharmacokinetics.

## 1. Affinity Chromatography

Affinity chromatography encompasses a great range of separation types (Goffe et al., 1988; Hrinda et al., 1989; Quinlan et al., 1992; Chadha et al., 1979; Hochuli et al., 1988; Maisano et al., 1989; Arnold, 1991; Sulkowski, 1985; Shadle et al., 1992; Knight and Fahey, 1981; Holmberg et al., 1976; Duffy et al., 1989; Narayanan and Crane, 1990; Hochuli, 1988; Ezzedine et al., 1992; Dean et al., 1985). From early use in research the field has burgeoned and many different chemistries are now available to bind various ligands covalently to numerous types of solid supports that range from silica

**TABLE 4**  Types of Chromatography and the Nature of the Separation Mechanisms[a]

| Name | Principle | Separation according to |
|---|---|---|
| Adsorption chromatography | Surface binding | Molecular structure |
| Ion-exchange chromatography | Ionic binding | Surface charge |
| Gel filtration | Steric exclusion | Molecular size and shape |
| Affinity chromatography | Biospecific adsorption and desorption | Molecular structure |
| Hydrophobic interaction | Hydrophobic complex formation in aqueous solvent | Hydrophobicity and hydrophobic patches |
| Covalent chromatography | Covalent binding (S—S) | Functional groups |
| Metal chelate chromatography | Coordination complex formation | Complex formation with transition metals (His, Cys) |
| Reversed-phase liquid chromatography | Hydrophobic complex formation in organic solvent | Hydrophobicity |

[a]Modified from Jungbauer, 1993.

to polymeric beads to membranes. The principle of affinity chromatography is to bind the product selectively through its specific epitope or binding site to a ligand which is covalently coupled to a chromatographic resin. Elution is achieved either by changing physical conditions such as pH or ionic strength, or by selective elution with a soluble ligand. The main advantages of affinity chromatography are its high selectivity and high capacity. Purification factors of 1,000 to 10,000-fold have been achieved using affinity chromatography in research, and it has moved to the process laboratory since the advent of affordable affinity resins with high cleanability, fast flow rates, and well-characterized ligands.

*Group-specific* affinity chromatography employs an affinity ligand that is recognized by a class of proteins. For example, protein A is specific for many animal species of IgG (Duffy et al., 1989), and triazine dyes will specifically bind to a number of proteins (Knight and Fahey, 1981). In a fermentation broth, many species might be expected to bind to such supports. Specificity is conferred by adjusting the chromatography conditions to achieve selective elution, for example by using shallow gradients or buffer additives. *Type-specific* affinity chromatography is much more customized to the product protein, and may need to be prepared by the user (Narayanan and Crane, 1990; Ezzedine et al., 1993; Hrinda et al., 1989). For example, an antibody that is specific for the product can be covalently coupled to a chromatographic support and used to purify that product. Both of these types of affinity chromatography are described in further detail below.

*a. Group-Specific Affinity Chromatography* Immobilized metal chelate affinity chromatography (IMAC), is a form of group-specific affinity chromatography that has been used to purify interferon-gamma and human growth hormone (Chadha et al., 1979; Maisano, 1989) at research and process scale, as well as numerous other proteins (Arnold, 1991; Shadle et al., 1992). The support has covalently bound chelating agent, to which metal ions (usually copper, zinc, or nickel) bind very tightly. After being "charged" with metal ions and equilibrated in a high ionic strength buffer that inhibits ion exchange interactions, the support binds proteins selectively through His and perhaps Cys residues (Hochuli, 1988; Maisano et al., 1988; Chadha et al., 1979; Shadle et al., 1992). The protein is eluted along with significant levels of leached metal ion by using an analog such as His, imidazole, by changing pH, or by stripping the column of metal ions using an agent such as EDTA.

IMAC separations have several advantages which are illustrated by the interferon example: The selectivity is excellent, the column has a high protein capacity, the chromatography is easy to scale up, and the resins are cleanable with sodium hydroxide. Regulatory concerns about the leaching of the metal ion are straightforward to address, since sensitive assays for metal ions exist. In fact, IMAC has been coupled with fusion protein technology to engineer a hexa-His metal binding site into a target protein, thereby facilitating its purification (Hochuli et al., 1988).

IMAC also has disadvantages, which should be carefully considered for each new protein product. The use of metal ions, particularly copper (II) or vanadium (II), introduces a catalyst for oxidation, and recently it has been reported that some proteins are damaged during chromatography (Quinlan et al., 1992; Shadle et al., 1992). The use of IMAC in purifying basic fibroblast growth factor generated a host of species which included protein oxidized at Cys residues, S-S bonded aggregates, and product fragments that resulted from peptide bonds cleaved as a result of cysteic acid formation (Shadle et al., 1992). Albumin was similarly shown to be altered by its exposure to vanadate ion on an IMAC column (Quinlan et al., 1992). Many residues on a protein can be damaged by oxidation, including disulfide bonds, His, Trp, and Met residues (Gracy, 1992; Fontana and Toniolo, 1976). The use of zinc and other ions which are less oxidizing, the exclusion of oxygen from column buffers, or the addition of protective agents may help to prevent the formation of such product degradants (J. Oeswein, 1994, personal communication). These types of changes are best detected in process intermediates by using peptide mapping and/or mass spectrometry. An IMAC step should be compared carefully to other separations to assure that the product quality is not compromised.

   *b. Protein A Affinity Chromatography*   A second group specific absorbent that has application in research and process purification as well as analytically is protein A (or G) affinity chromatography (Duffy et al., 1989; Narayanan and Crane, 1990; Dean et al., 1985). Protein A binds to immunoglobulin G (IgG) of several species in the heavy chain Fc region with high selectivity. Covalently coupling protein A to agarose, silica, controlled pore glass, and various polymeric resins has resulted in chromatography resins with high capacity and selectivity for IgG from several mammalian species. Most of the chromatography resins have excellent cleanability, flow properties, and dynamic capacities and can be used for hundreds of chromatography cycles. Protein A itself is chemically stable to low pH, brief chaotrope exposure, and to some organic solvents, aiding in the sanitization of protein A affinity columns. Recent improvements in resin chemistries and the inherent stability of protein A have permitted the use of protein A affinity chromatography as a capture step or a downstream purification step in process purification of many antibody products (Duffy et al., 1989; Narayanan and Crane, 1990; Hochuli, 1988; Ezzedine et al., 1993; Dean et al., 1985; Johnson et al., 1994). The main advantage of using such a protein A affinity column is that it permits capture from a dilute and impure feedstream with purification to high purity in a single step. This will, in general, reduce the size of downstream columns and reduce the number of unit operations required.

   As with other protein ligand affinity separations, the disadvantage of using protein A chromatography lies in regulatory concerns that affect QC and process validation. The storage and cleaning of the resin and maintenance of microbial cleanliness are more difficult than with resins that can be cleaned and stored with sodium hydroxide. It may be necessary to develop assays for leached protein A, validate clearance of protein A in the purification process, and even set a specification for allowable levels of protein A residuals in the final product. The effect of column aging on the purity and stability of the product should be carefully evaluated. The source of the protein A used to prepare the column, and methods used to purify the protein A are also important to consider, since these may add new regulatory concerns.

A third group specific absorbent that has been used to purify native and recombinant proteins is dye ligand chromatography. This method uses a dye which binds selectively to certain classes of proteins. Dye ligand chromatography has been used, for example, in research purification of human fibroblast interferon (Knight and Fahey, 1981). A major concern of dye ligand chromatography for manufacturing is the leaching of the dye from the resin. Because many dyes are chemically heterogeneous, are closely related to carcinogenic compounds, and have not been well studied for tumorigenicity in animals, serious safety concerns must be carefully addressed before they can be used in a GMP process.

*c. Type-Specific Affinity Chromatography* Type-specific affinity chromatography has been used in purification of human blood factors VIII and IX from blood plasma, as well as recombinant urokinase, by covalently coupling a monoclonal antibody which specifically binds to the product to a chromatographic support and using it to purify the product (Hrinda et al., 1989; Hochuli, 1988; Ezzedine et al., 1993). These type specific affinity chromatography separations share the advantages of high capacity, speed, and high selectivity. A high purification was achieved with an easy step gradient elution method from a complex starting material. The methods have been scaled up successfully and some are in use for marketed products. Several of these products also used group-specific affinity chromatography in the same process, for example factor VIII purification included affinity chromatography on His (Hrinda et al., 1989) while urokinase was purified from *E. coli* using a benzamidine affinity column which bound to the active site of the enzyme (Holmberg et al., 1976).

The disadvantages of using a type specific ligand such as a monoclonal antibody include safety and economic issues. If a monoclonal antibody derived from mouse ascites is used as a ligand on an affinity column, it may be contaminated with host cell proteins, viruses, DNA, or endotoxin. The affinity ligand therefore may require its own validated manufacturing process and complete characterization in order to be used as a reagent to purify a therapeutic protein in a GMP process. If prepared from mouse ascites, viral safety must be evaluated on every batch, which is very costly.

A second safety issue is the leaching of the ligand from the resin, which occurs at a detectable level that varies with the type of coupling chemistry used, the age of the column, and the properties of both buffers (pH may catalyze hydrolysis) and the feed (a protease may cleave ligand off the column). In general, the use of a type specific affinity column necessitates the development of a high sensitivity immunologic assay for ligand which can detect ppm levels in final product. This testing is then added to the release testing of final product. In the manufacturing plant, assays should be in place to monitor the performance of the affinity column, and process validation should demonstrate adequate performance across the column lifetime.

Economic issues become important to consider when choosing a type specific affinity separation for the manufacturing plant. Addressing the safety concerns above may be very costly for some affinity ligands, lessening the relative advantages of affinity chromatography as compared to traditional methods (ion exchange, for example). The supply of the ligand may be too limited to supply a manufacturing plant, and indeed, it may be the most costly raw material used in a process. With the human blood factors, the size of the market for the product, the ability to reuse the columns for hundreds of cycles, the need to rapidly purify product away from proteases, and regulatory requirements for purification steps with high viral clearance justified these regulatory and manufacturing costs.

## 2. Ion Exchange Chromatography

Ion exchange chromatography (IEX) is one of the oldest chromatographic methods, and is used in most purification schemes, sometimes more than once. Anion exchange, in which negatively charged components are bound, and cation exchange, in which positively charged components are bound, are both in common use. Both weak and strong functional groups are useful in purifying proteins from contaminants, and may have different properties that depend on the functional group and the chromatographic support (Jungbauer, 1993). Ion exchange chromatography may remove DNA, endotoxin, and host cell proteins from a protein product. Both cation and anion exchange chromatography may be able to separate product from such product variants as altered sialylation, deamidated forms, and truncated sequences, provided they differ in charge. Table 5 compares the major modes of chromatography for their primary use, advantages and disadvantages.

IEX is an effective capture step because it can reduce volume and effect significant purification, usually with simple step gradient elution. It can also be used as a polishing step to separate closely related product variants, with either step or linear gradient elution. After product binds to the column at low ionic strength, an IEX separation can be developed by elution with increasing ionic strength, changes in pH, or both. Most IEX columns will tolerate additives such as urea or methanol which may enhance the solubility of the feedstream or inhibit hydrophobic interactions with the support or other proteins. IEX columns have moderate to high

TABLE 5 Comparison of Purification/Separation Chromatography Types

| Process step | Advantages/disadvantages (+/–) |
|---|---|
| Ion exchange chromatography | + Selective, powerful for capture step; high capacity |
| | + Scales up well; cleanable; well understood |
| Anion exchange | – Binds nucleic acids |
| Cation exchange | + Does not bind DNA, endotoxins; high removal capability |
| | – May require pH changes where proteins are not stable |
| Size exclusion chromatography | + High yield, low shear, minimizes degradation of product |
| | – Dilutes product into desired buffer; rate limiting process step |
| | – High cost and limited scalability |
| | – Lower in resolution than other modes |
| | – Low capacity for volume and mass |
| Affinity chromatography | + Highly selective, high capacity |
| Immobilized metal affinity chelate | + Selective, scales up well, high capacity |
| | – May degrade some proteins; toxic waste stream |
| Nonprotein ligand | + Selective; cleanable; scales up well; usually moderate in cost |
| | – Leached ligand may have regulatory or safety issues |
| Protein ligand | + Highly selective; high capacity, scalable |
| | – Leaching of ligand may have regulatory or safety issues |
| | – Costly resin and ligand; potential supply issues; second manufacturing process needed (for ligand) |
| | – Cleaning more difficult than nonprotein ligand |
| Hydrophobic interaction chromatography | + Selective, powerful purification step; links well to ion exchange; cleanable |
| | ± Moderate capacity |
| | – Product or contaminants may precipitate by salting-out (Halenbeck et al., 1988; Wetzel, 1992) |

*(continued)*

TABLE 5   Continued

| Process step | Advantages/disadvantages (+/−) |
|---|---|
| Covalent chromatography | + Selective, scales up well<br>− May be costly<br>− Not compatible with reducing agents<br>− May cause unwanted disulfide exchanges |
| Reversed phase chromatography | + Selective, powerful; useful for peptides and proteins<br>+ Links well to drying step<br>− HPLC equipment is costly, and separation may not scale up well due to problems with packing large columns and reproducibility of gradients<br>− Waste disposal issues to consider |

capacity, are not sensitive to the volume of feed loaded, and can be easily cleaned. The main disadvantage of IEX is that the product is usually eluted in a high salt buffer, thus needing additional handling before it is ready for formulation.

## 3. Hydrophobic Interaction Chromatography

Hydrophobic interaction chromatography is performed when a column is equilibrated in an aqueous solvent and components bind via hydrophobic interactions (Melander et al., 1984). In general, components are bound at high salt and elute at low salt or when an organic solvent is introduced. The most common functional groups used as ligands on HIC columns are phenyl, butyl, and ether. These ligands differ in how strongly they adsorb proteins. A strongly hydrophobic resin, such as a butyl resin, may bind protein at high capacity, yet may also have reduced yield because of irreversible adsorption. A weakly hydrophobic resin (ether, for instance) will have higher yields for the same protein but may have lower resolution and capacity. These properties are fine-tuned by careful choice of salt composition, concentration, temperature, and pH.

HIC is an effective purification method that links well with ion exchange steps because of its inverted salt concentration requirements (Jungbauer, 1993). HIC is a high resolution technique and has been used to separate proteins from host cell proteins, to fractionate impurities such as oxidized or misfolded product, alternate glycoforms, and to remove DNA, endotoxin, and other host contaminants (Sofer and Nystrom, 1989; Jung-bauer, 1993; Halenbeck et al., 1988). HIC steps, once developed, are excellent polishing steps for many products. The high resolution, unique selectivity, and use of relatively innocuous aqueous buffers make this type of chromatography desirable.

Compared to ion exchange, HIC is more limited as to where it will fit in a purification scheme. The product or contaminants may be prone to aggregate or precipitate in buffers used in HIC (Halenbeck et al., 1988), reducing yield and purity by coprecipitating with product. Thus, HIC is not the first choice for capture steps. The high concentrations and unusual salts that are used in the chromatography may mean that its eluate requires further handling to buffer exchange it to a formulation ready state. Furthermore, the chromatography is very sensitive to temperature changes. Small changes in temperature may result in a large change in retention time, peak shape, and resolution of product from contaminants and impurities. Careful process development is required to define the proper limits of these and other critical process parameters.

**Example. Purification of recombinant human macrophage colony-stimulating factor (M-CSF) expressed in CV-1 cells.** HIC was used in two serial steps to purify recombinant M-CSF expressed in CV-1 mammalian cells (Halenbeck et al., 1988). It was determined that host contaminants would precipitate at the high salt concentration needed to bind product to the column. Therefore, the first HIC step was performed as a pass-through step at low ionic strength (about 0.1 molar) to remove hydrophobic host contaminants. The ionic strength was then increased to over 1 molar, and the feed applied to the same type of HIC column. At this salt concentration, the product bound and purification was then carried out by gradient elution.

## 4. Reversed Phase Chromatography

Reversed phase liquid chromatography is a powerful method for the purification of peptides and of proteins of small and moderate molecular weight (Jungbauer, 1993; Bishop et al., 1980; Hilaireau et al., 1990). RP-LC separations are often sensitive to minor sequence changes, such as occur in synthetic peptides when single amino acids are deleted during synthesis (Rivier et al., 1984), or when deamidation, oxidation, or other degradation reactions occur. At process scale, RP-LC or HPLC may resolve these product variants partially or completely depending on the structure of the product.

At analytical scale, RP-HPLC resolved single-amino acid changes that had been introduced by site-directed mutagenesis in rIL-2 (Kunitani et al., 1986b) and in insulins from various animal species (Kroeff et al., 1989). RP-HPLC has also separated intact polypeptide from polypeptide containing a single Met sulfoxide residue (Kunitani et al., 1986a), folded from misfolded proteins (Kroeff et al., 1989), and numerous other degradants. At process scale the separation of these species is usually much lower because the process chromatography resins have a larger particle diameter which reduces resolution, and more protein is loaded per unit volume. Often a larger yield loss must be taken to achieve the same degree of purification, if quantitative removal of such an impurity is required. RP-HPLC also varies in its ability to resolve glycosylated from non-glycosylated, or carbohydrate chain variants, of glycoproteins.

Because of its high selectivity, RP-HPLC is commonly used as a method of product analysis. It has also been used in large scale purification, for example insulin (Kroeff et al., 1989), GM-CSF (Urdal et al., 1984), and IL-2 (Kunitani et al., 1986a, 1986b). RP-LC has several advantages, chief among them its power as a separations technique. Also, by providing product in a volatile buffer, RP may make subsequent lyophilization easier.

Several issues arise when RP-LC is scaled up for manufacturing. One, if a high-performance separation is required, the pressure rated purification equipment is considerably more costly to design, build, and maintain. Two, achieving equivalent packing quality of columns and reproducibility of making the gradient is difficult and costly when the chromatography is scaled up to manufacturing scale. Three, the solvents used for reversed phase chromatography may require facility modifications such as explosion proof rooms, walk-in fume hoods, and toxic waste disposal. Solvent recycling is usually necessary for economic reasons once at manufacturing scale. Last, some ion pairing agents such as TFA bind to proteins and may carry through to the final product at unacceptable levels.

## 5. Size Exclusion Chromatography

Size exclusion chromatography (SEC) is commonly used as a purification method in research and early process development, but has severe limitations when applied to protein purification at process or manufacturing scale. In general, the resolving power is significantly lower than for methods such as ion exchange or HIC. Other issues are that the flow rates are very low, product is diluted during purification, and the capacity for protein and injected volume per run is very low (5 to 20% of column volume depending on the separation). The volume limitation may add another unit operation for concentration both before and after the SEC step. Together, these limitations make the SEC column the rate-limiting one of a process. The chromatography is sensitive to the quality of packing of the column and to wall effects, which means that performance may not scale up. To maintain resolution and minimize pressure problems, large columns must be poured in multiple shorter modules in addition to being cycled multiple times per batch.

Where SEC has much more utility, and competes with alternatives, is in buffer exchange steps where high resolution is not needed. SEC in this application should be compared to tangential flow ultrafiltration, which it may replace if the product is shear sensitive. For greater efficiency, the sample can be applied in a "boxcar" mode; i.e., before the included volume of the previous injection is eluted. Recombinant human insulin purification utilized SEC steps for buffer exchanges (Kroeff et al., 1989).

## D. Process Changes

During clinical development, process changes will be made. For example, the process may be scaled up, a new unit operation may be added, or changes may be made to operating conditions or pooling criteria. Pooling of appropriate fractions from a chromatography column must be carefully defined in order to assure consistency across scale-up into new equipment. In-process testing verifies that pooling was done properly and the data also support process validation and monitoring.

Process changes may or may not affect the measured attributes of final product such as final purity, however, they may affect stability or potency. For this reason, the formulation scientist needs to be aware of upstream changes and perform appropriate studies to determine whether final product and its properties have been affected. In addition, process chromatography steps may suggest stability indicating methods and may generate side fractions enriched in impurities that are useful in developing and validating assays.

## V. UNIT OPERATIONS THAT ALTER THE PRODUCT

A number of nonchromatographic unit operations are commonly used in a purification process to achieve goals other than purification, concentration, or clarification. Among these are protein refolding, viral inactivation, and chemical or enzymatic cleavage of fusion proteins.

## A. Protein Refolding

A recombinant protein expressed in bacteria may be isolated in a form which does not possess biological activity because it is aggregated, precipitated, and/or in a reduced state. These proteins require refolding of the polypeptide chain, and may also require oxidation to form the correct intra- and interchain disulfide bonds (Tsuji et al., 1987; Kunitani et al., 1986b; Kunitani et al., 1986a; Shaked and Wolfe, 1985; Winkler et al., 1985; Hanna et al., 1991; Gilligan et al., 1987). Refolding of recombinant proteins is a complex field, and has been reviewed by many authors (Marston, 1986; Uhlen et al., 1992; Kane and Hartley, 1988; Seely and Young, 1991; Wetzel, 1992; Valax and Georgiou, 1991; Babbitt et al., 1990). Refolding conditions often have to be empirically determined for a particular amino acid sequence, and the sequence rules governing refolding requirements are poorly understood. The outcome of a refolding experiment is also affected by the presence of contaminants and impurities.

The process scientist must find appropriate chemical and physical conditions in vitro to promote proper refolding and oxidation of the correct disulfide bonds, while suppressing other chemical reactions that result in degradation or denaturation. A recombinant protein placed under conditions which promote refolding may have available to it many competing pathways. It may form noncovalent aggregates, it may form disulfide bonded aggregates, or it may precipitate. Chemical degradation, such as oxidation, deamidation, or carbamylation may occur. The protein may have several isoforms which are intermediates along the pathway to properly folded product. In the case of tryptophane synthetase, these intermediates were shown to vary in solubility and propensity to aggregate (London et al., 1974). These differences in stability and solubility of folding intermediates explained why the recovery of active material varied unexpectedly with chaotrope concentration. This suggests that in optimizing the pH, ionic strength, and buffer composition of a refolding unit operation, the scientist may be determining conditions which favor the stability and solubilities of several different isoforms (Wetzel, 1992). The process scientist seeks to find a balance empirically between kinetics and thermodynamics of competing pathways which will allow the recovery of correctly folded and oxidized protein and the removal of any misfolded, degraded, and denatured species.

### 1. Refolding of S-Sulfonated Protein

Formation of a disulfide bond restricts the movement of the polypeptide chain and limits the possible conformations it may have (Glaser et al., 1982).

Such restriction can help to favor correct refolding of a protein, or it may "trap" the protein in an incorrect conformation, from which it is impossible to form a correctly folded, active product (Morehead et al., 1984). Refolding the peptide backbone before allowing the formation of disulfide bonds allows better control over refolding and disulfide bonding for many proteins.

A common strategy is to block free sulfhydryl groups with an agent such as sodium tetrathionate to form the sulfonated (Cys-S-SO$_3$) polypeptide. The S-sulfonation method involves covalent addition of SO$_3^-$ groups to free sulfhydryls on the protein or peptide before refolding. Sulfonation changes the charge of the polypeptide, thereby affecting its solubility, stability, and its retention on ion exchange and reversed phase chromatography.

This refolding strategy was used to refold recombinant human alpha-interferon A which had been expressed in *E. coli*, isolated as a soluble intracellular protein at research scale, and highly purified in a monomeric state (Wetzel, 1992; Morehead et al., 1984). The purified monomeric interferon was reacted with sodium tetrathionate and sodium sulfite. After sulfonation of free thiol groups was completed, the protein was buffer exchanged to remove reaction products, further purified, and permitted to refold. Refolding was carried out in the presence of an empirically determined concentration of a chaotrope (guanidine HCl). Other conditions such as low temperature, low protein concentration, and careful choice of buffering and salts improved the yield of properly folded product.

When refolding was completed, the sulfonation was reversed by a thioldisulfide exchange step. Glutathione (oxidized GSSG and reduced GSH, added at an empirically determined ratio) was added. The reduced form of glutathione (GSH) rapidly displaces the protein-bound S-sulfonate to form mixed disulfides between GSH and the protein cysteine residues. These intermediates undergo further thiol-disulfide exchange, facilitated by GSH and GSSG, to encourage formation of the protein disulfide bonds. Further purification removed the process chemicals and any product which had failed to refold and form disulfide bonds correctly.

A similar method was used to refold proinsulin at manufacturing scale (Chance et al., 1981). Sulfonated proinsulin was partially purified in the presence of a chaotrope (urea). Refolding was induced by reducing the chaotrope concentration and incubating at low temperature in a defined buffer. The sulfonated, refolded protein was converted to native conformation proinsulin by adding beta-mercaptoethanol under controlled conditions which promoted formation of correct disulfide bonds (Kroeff et al., 1989; Kemmler et al., 1971; Chance et al., 1981). Further purification and enzymatic conversion to insulin completed the process (Kemmler et al., 1971; Kroeff et al., 1989).

In performing refolding and oxidation by this approach, the process scientist is concerned about the purity of the process chemicals used and about damage to the product during the long incubations. The sulfonation reaction does generate free radicals, and sensitive residues such as Met, Trp, and His may be damaged. Chelating agents such as EDTA may protect against metal-catalyzed oxidation. Sulfonation can protect the protein against forming incorrect disulfide bonds during early refolding, and the sulfonate group is a good leaving group. In contrast, glutathione and cysteine may stay bound to proteins and be difficult to detect (Glaser et al., 1982).

## 2. Combined Refolding/Oxidation Reactions

Many proteins can be refolded concurrently with formation of any disulfide bonds, whether the S–S bonds are intra- or interchain. Exposure to air during purification will allow spontaneous refolding to occur. A metal ion such as copper (II) can be added to catalyze the process. Air oxidation may be difficult to scale up because surface to volume ratios change, and it may occur very slowly. It also generates peroxides, which can oxidize sensitive amino acid side groups (Wetzel, 1992).

Peroxide generation is avoided by a second method, in which refolding is catalyzed by the use of thiol–disulfide exchange. Adding a thiol-containing process chemical such as glutathione, cysteine, or beta-mercaptoethanol to the denatured, reduced product promotes an equilibrium disulfide exchange reaction (Tsuji et al., 1987; Kunitani et al., 1986a; Gilligan et al., 1987). Usually a chaotrope must be added to prevent aggregation and precipitation of the starting material and folding intermediates. Chaotrope at low concentrations also enhances yield, probably by increasing the flexibility of the polypeptide backbone and permitting more frequent interconversions in conformation.

In most proteins, the properly folded and S–S bonded isoform is thermodynamically favored, so that after incubation, active product can be recovered at high yield. The properly refolded protein may be unique in its solubility, and in retention time on reversed phase HPLC (Wetzel, 1992). It may differ in SDS-PAGE migration, and exhibit other recognizable physical properties. Yield of active product depends on the conditions used, and depends on the kinetics of the desired refolding pathway as compared to competing pathways of degradation, denaturation, aggregation, or precipitation.

One concern in using thiol–disulfide exchange in a manufacturing plant is the high cost of such thioldisulfide exchange reagents as glutathione, and a chaotrope such as guanidine HCl. Minimizing the volume of the refolding reaction helps to reduce these costs, yet may be at odds with the need to inhibit aggregation and precipitation. Similarly, low temperature and low protein concentration, conditions which improve the yield and purity of refolded active product, adversely affect the cost, scalability, and throughput of the unit operation. During early development, the scientist usually starts with extremely favorable conditions which give a high yield. During optimization, these conditions are varied to improve the practicality and economics of the operation.

## 3. Product Degradation During Refolding

Refolding steps usually take a long time compared to other processing steps. A product may degrade significantly during refolding, especially if the conditions are not favorable for its stability. Aggregation and precipitation may occur, especially at high protein concentrations, and may be reversible or irreversible (Wetzel, 1992). The use of chaotropes or detergents assists in preventing these types of reactions, and also improves the rate of refolding by allowing the polypeptide chain more movement.

Conversely, buffer additives may promote degradation. For example, the chaotrope may expose amino acid residues to aqueous conditions and increase the rates of deamidation or oxidation. If urea is being used as a chaotrope to prevent aggregation and misfolding, free amino groups may be carbamylated by cyanate, a breakdown product of urea. Carbamylation is of concern since it has been shown to inactivate some (but not all) proteins and to target other proteins for turnover in vivo (Hood et al.,1 977; Gracy, 1992). If disulfide bonding is being promoted by the use of a high pH, carbamylation may be rapid in some proteins (Hood et al., 1989). Carbamylation can be inhibited by purifying urea using ion exchange chromatography just prior to use, and by adding amino-group-containing buffers such as glycyl glycine as competitive substrates (Wetzel, 1992). Low temperature, short exposure time, and avoidance of high pH also help, but some products are so rapidly modified that it may be necessary to use other chaotropes. Conductivity is an easy in-process assay which can be used to monitor the generation of cyanate ion in urea solutions (Wetzel, 1992). Similarly, copper ion has been used as a catalyst to promote disulfide bond formation, but may be associated with damage to other residues. For example, formation of Met sulfoxide occurred in refolding recombinant human IL-2 made in *E. coli* (Shaked and Wolfe, 1985).

Deamidation is a sequence-dependent event which is catalyzed by the high pH conditions that promote interchain disulfide bond formation. In deamidation, the conversion of asparagine (or less commonly glutamine) residues to aspartic acid or to isoaspartate (isoasp) may occur. Isoasp formation can result in cleavage of the peptide bond and fragmentation of the protein (Johnson and Aswad, 1990; Johnson et al., 1989). Deamidation is usually detected by a small shift in the isoelectric point of the protein (Righetti et al., 1989). Since many proteins have multiple Asn and Gln residues, numerous species may be generated. Deamidated proteins may be fully bioactive with identical pharmacokinetics, or may have decreased potency, reduced in vivo circulation half-lives, or even be incapable of forming proper quaternary structures (Johnson et al., 1989). The important goal of the process scientist is therefore to generate a consistent ratio of the different species, or to minimize generation of deamidated species. Since deamidation is often not preventable, it is most common to characterize a mixture for its biological and toxicological activities in vitro and in vivo during process development.

Any other type of degradation to which the protein is prone may occur during refolding steps as during the rest of the process. For example, *E. coli* expressed creatinine kinase was degraded by a contaminating protease which was activated only during the refolding step (Babbitt et al., 1990). Careful in-process characterization is needed to determine the point in the process where such degradation occurs, and its mechanism.

## B. Viral Inactivation Treatments

When recombinant proteins are expressed in mammalian cell lines, current regulatory guidelines require that viral clearance (inactivation plus removal) be evaluated (Points to Consider, 1994; Points to Consider, 1993; CPMP Draft Guidelines, 1993). The primary viral safety concern is the potential presence of type A and C retroviruses derived from the cell line. Retroviruslike particles can be observed by EM analysis of most mammalian cell lines (Mariani and Tarditi, 1992; White et al., 1991; Lubiniecki et al., 1990; Lubiniecki et al., 1988; Levine et al., 1992), yet have not been shown to have viral infectivity (Lubiniecki, 1988). To be prudent, a worst-case risk assessment is made by assuming that the number of viruslike particles is proportional to the risk of viral disease in human patients. Risk is estimated by collecting conditioned media at the end of the cell culture, concentrating it, then analyzing by electron microscopy. The number of viruslike particles per dose of final product is calculated mathematically and used as a quantitative estimate of potential risk.

The process purification scientist then has the task of demonstrating that the purification process is capable of clearing a significant excess of a model retrovirus in a spike challenge experiment. Clearance (removal plus inactivation) of a model retrovirus is evaluated in a scaled-down purification process, and should provide a safety factor of excess clearance of several logs over the estimated number of viruslike particles (White et al., 1991; Lubiniecki et al., 1990; Levine et al., 1992). Treatments that have been used to inactivate retroviruses include extremes of pH, addition of chaotropes, elevated temperature of 55°C or more, or addition of surfactants, organic solvents, or other agents with virocidal properties (White et al., 1991; Lubiniecki et al., 1990). Removal is achievable by many modes of chromatography, but the amount of clearance depends on the chromatographic conditions used and cannot be predicted in advance. Clearance is defined as the log of the amount of model virus detected in the product eluate fraction divided by the amount of virus loaded onto the column, and includes removal as well as inactivation. Evaluation of viral safety usually also includes a similar evaluation of the removal of other model viruses representing enveloped and nonenveloped DNA and RNA virus types. However, no estimation of risk can be performed which is comparable to counting the retroviruslike particles.

Viral inactivation treatments are of concern to the protein chemist, because they may induce significant product denaturation or degradation in some proteins. Aggregation, deamidation, and oxidation are perhaps the most common effects but other types of degradation are possible and vary with the sequence and structure of the protein. For example, high pH treatment may racemize prolines, isomerize asparagines, and cleave peptide bonds. Some viral inactivation treatments may also add new process chemicals to the product. Each of these aspects may affect product stability and affect the types of methods required for testing of final or bulk product. It is essential to perform adequate stability studies on process intermediates using the viral inactivation conditions for extended time. The formulation scientist can assist the process development scientist by providing stability-indicating methods early in development.

## C. Cleavage of Fusion Proteins

Because of the biological limitations of recombinant expression systems, some protein and most peptide products may be expressed recombinantly as fusion proteins with a domain of another protein (Marston, 1986; Uhlen et al., 1992; Gilligan et al., 1987; Dahlman et al., 1989; Saito et al., 1987;

Forsberg et al., 1992). The fusion protein may be needed to allow expression, prevent turnover of the recombinant protein within the cell, allow it to be secreted, or change the solubilization properties to aid in purification (reviewed in Uhlen et al., 1992). For example, proinsulin was expressed as a fusion protein with the N-terminal sequence of a bacterial protein, and then required cleavage at the Met residue with cyanogen bromide, refolding, and enzymatic treatments to convert it to insulin (Goeddel et al., 1979; Williams et al., 1982; Kroeff et al., 1989; Kemmler et al., 1981; Chance et al., 1981).

Cleavage of a recombinantly expressed fusion protein to release a peptide product has been done using various chemical and enzymatic treatments, as summarized in Table 6. For example, a lung surfactant peptide was expressed as a fusion protein with a domain of chloramphenicol acetyltransferase, a bacterial protein, and then was cleaved at Asn-Gly residues using hydroxylamine at high pH (Benson et al., 1993; Bornstein and Balian, 1976). The fusion partner caused the product to be localized in inclusion bodies. Such a chemical method of cleaving the desired product has less utility with protein products than with peptide products, since chemical selectivities are usually determined by only one to two amino acids in the sequence, such as may occur multiple times in a longer polypeptide sequence (Uhlen et al., 1992).

Many proteins have been expressed with an affinity domain fused on to improve purification properties. For example, the addition of a six amino acid long His repeat to either end of a target protein facilitates generic purification by IMAC chromatography, and may improve the solubility of some proteins (Hochuli et al., 1988). Similarly, a DNA-binding domain of the glucocorticoid receptor was expressed by fusing it with a domain of protein A, which was used for purification and then cleaved off using chymotrypsin (Dahlman et al., 1989). Several proteins, including the cystic fibrosis transmembrane conductance regulator, were expressed in insect cells as a fusion with glutathione-S-transferase, which provided an affinity domain useful for purification. The fusion domain was then cleaved off by using an engineered thrombin or factor IX site (Nagai et al., 1985). Among the most useful enzymatic methods in use today are enterokinase, engineered subtilisin, and factor Xa (Uhlen et al., 1992; Forsberg et al., 1992; Carter, 1990). Each of these enzymes possesses a sequence specificity which allows for high selectivity on even large protein products, thus minimizing the yield losses and generation of impurities during the cleavage reaction. In addition, each of these enzymes can be economically produced at high purity for use at larger scale.

**TABLE 6**  Cleavage Sites Engineered into Fusion Proteins in *E. coli* [a]

| Sequence recognized | Cleavage effector | Literature reference |
| --- | --- | --- |
| –Asp–↓–Pro– | Acid pH | Szoka et al. (1986) |
| –Met–↓–Xaa– | CNBr | Goeddel et al. (1979) |
| –Arg–↓–Xaa– and –Lys–↓–Xaa | Trypsin | Shine et al. (1980) |
| –Ile–Glu–Gly–Arg–↓–Xaa | Factor Xa | Nagai et al. (1985) |
| –Pro–Xaa–↓–Gly–Pro–Yaa–↓ | Collagenase | Lee & Ullrich (1984) |
| –Arg–↓–Xaa | Clostripain | Bennett et al. (1984) |
| Asn–↓–Gly | Hydroxylamine | Canova-Davis et al. (1992) |
| R–↓–X–COOH | Carboxypeptidase B | Kemmler et al. (1971) |

[a] Modified from Marston, 1986.

Cleavage of a fusion protein is a unit operation that is customized to the product, the fusion partner, and the mode of expression. Chemical cleavage methods include, for example, CNBr to cleave selectively after Met residues, and hydroxylamine to cleave at Asn-Gly sequences (Benson et al., 1993; Bornstein and Balian, 1976; Means and Feeney, 1971; see Table 6). These and other chemical methods, however, are not completely selective, and can result in low yields, side reactions with product or contaminants, and degradation of the product under the harsh conditions needed to drive the chemical reaction (Uhlen et al., 1992). For example, Canova-Davis et al. showed that hydroxylamine cleavage used in making recombinant human IGF-1 (insulinlike growth factor 1) generated a host of product variants which were detectable on ion exchange HPLC and included hydroxamates of Asn and Gln as well as Met sulfoxide (Canova-Davis et al., 1992). Another possible issue with chemical cleavage reactions is that cleavage may occur at alternate sites which are chemically similar to the intended ones (Means and Feeney, 1971).

In contrast to chemical cleavage, enzymatic reactions can be carried out under more physiological conditions and are highly selective for certain amino acid sequences. Enzymatic application has been limited for the most part by the lack of availability of large quantities of highly purified enzymes. Trypsin, chymotrypsin, carboxypeptidase B, and factor X have all been used with success (Gilligan et al., 1987; Kemmler et al., 1971; Ueno and Morihara, 1989; see also Table 6). More recent choices of enzymes include enterokinase and an engineered version of subtilisin (reviewed in Carter P,

1990; Forsberg et al., 1991, 1992). The high cost of some enzymes can be mitigated by utilizing covalent coupling to a chromatographic resin and validating repeat use. Of course, active enzyme may leach into the product and should be evaluated as a potential process chemical contaminant.

## VI.  DESIGN OF PURIFICATION PROCESS FLOW

Optimizing a process involves arranging the unit operations in a sequence that minimizes extra handling between steps, positions volume reduction steps early in the process, and contains natural holding points where the process can be halted if necessary (Jungbauer, 1993). When repeated buffer exchanges are required to connect one unit operation to the next, the process is inefficient since such steps do not have purification power but add cost, time, and yield losses. The process should have the minimum number of operations necessary to achieve purity standards. In general, the most effective purification step (that which removes the most contaminants) should be first. This keeps downstream columns and equipment as small in size as possible, and helps to define the targets of downstream unit operations as polishing steps.

The order of steps may be profoundly affected by product stability issues. For example, tPA is susceptible to protease cleavage, making a primary goal of early purification that of separating cell derived proteases from the product (Arathoon et al., 1990). The use of an affinity column that may leach harmful process chemicals probably necessitates that it be placed upstream of a unit operation designed to remove the process chemicals (Levine et al., 1992). Placement of process steps also affects the ease of obtaining formulation-ready product from the last step. For example, placing an HIC column with its high salt buffers at the end of a sequence causes more difficulties in obtaining an isotonic bulk product than placing a size exclusion column last. Refolding step positioning has trade-offs: often, a protein can be refolded out of a relatively crude extract, allowing the subsequent process purification steps to be used to separate both contaminants and folded from misfolded product. However, the refolding yield may be somewhat lower out of the less pure starting material (Wetzel, 1992).

At the end of purification, the product is at an acceptable level of purity and homogeneity and is preferably in a formulation-ready buffer. If it is in the formulation buffer, it is ready for final filling or lyophilization; if not, further manipulation or addition of excipients is required. The product has now been exposed to a number of different buffer conditions and process chemicals. Some of these may be known to catalyze degradation. It may have already undergone significant degradation events, such as aggregation, deamidation, oxidation of Met residues, or mild proteolysis. Trace residues of proteases, metals, or other agents may or may not be present. Even at levels which cannot be directly detected, such agents may profoundly affect the stability of the protein. This is why it is highly desirable to evaluate the stability of more than one bulk lot of product early in development. Stability studies may be the only means by which the process scientist can determine whether the process is consistent and achieves adequate purity and homogeneity in each batch of the bulk product.

## VII. IN-PROCESS CHARACTERIZATION OF PROTEIN PRODUCTS

Biopharmaceuticals have special regulatory concerns which arise because of their chemical complexity and the means of production (Federici, 1994). The analytical approaches taken for monitoring the progress of the process reflect these concerns. Table 7 summarizes these concerns, which relate to the complexity of the product and the process. Because a protein can degrade by multiple means, its quality can only be assessed by a battery of assays. This is also true of many of the raw materials used in the cell culture or purification processes. Characterization of the product is performed both in-process, as a rapid check on the performance of the process, and after the process run is completed as a release test (see Chapter 6). Numerous reviews of analytical methods exist (terAvest et al., 1992; Rivier and McClintock, 1983; Federici and Garnick, 1991; Dinner and Lorenz, 1979;

TABLE 7 Concerns in the Quality Control of Biotechnology Products

Larger more complex molecules
Complicated manufacturing systems
Long production times
Pooling several fermentation runs
Raw material concerns
   Living organism
   Various raw materials used
   Special noncompendial requirements
Genetic stability

Reprinted from Federici, 1994.

Garnick et al., 1988; Jones and O'Connor, 1985). In this section, the use of the major modalities of assays for in-process monitoring will be discussed.

In-process assays may be performed to make go/no go decisions on proceeding with the batch (Sofer and Nystrom, 1989). Assays of this type are usually optimized for speed, and some may even be performed by production staff rather than in the QC lab. Protein concentration, LAL, and product concentration assays usually must be completed before a batch proceeds to the next process step. However, many in-process assays may be performed while the batch continues to move through the purification process (McEntire, 1994). For instance, it is not feasible to wait seven days for bioburden testing to read out. LAL assay is usually used as a surrogate for go/no go decision making, while bioburden results may result in later reprocessing, investigations, or extra testing. This latter type of in-process assay may be used to decide whether the batch is failed, or the information may be useful to document process control, qualify process improvements, or monitor the process. Assays such as SDS-PAGE, tests for process contaminants such as DNA, and immunoassays for host proteins may all be performed to gather information on the consistency of the process. These assays may not have rigid acceptance specifications placed on them, yet may be used to monitor performance of the equipment and facility.

The chief advantage of permitting the process to proceed is that an unstable process intermediate need not be held and allowed to degrade while testing is done. The main requirement of in-process assays is that they indicate the performance of the process and that they are predictive of final bulk or final product test results, which often do not read out for weeks or months after a batch is purified.

## A. In-Process Assays for Product

In-process assays usually include measurement of *product concentration* in feedstreams (Table 8). ELISA, HPLC, and bioactivity assays for product function have been most commonly used in this arena. To be useful as an in-process assay, such a method must be tolerant of various buffers and nonproduct contaminants which may interfere. The assay should also be performed quickly and results available in real time. The assay may be used to measure titer during cell culture and yield during purification, which aids in optimizing the process as well as monitoring a process for ongoing performance. Each unit operation usually is monitored for both yield and mass balance of product (i.e., amount recovered in all fractions compared to amount loaded). The latter is especially useful for cleaning validation,

TABLE 8   In-Process Assays Commonly Used for Recombinant Proteins

| Analysis | Methods used |
|---|---|
| Product concentration | ELISA, HPLC, affinity column, RIA |
| Total protein | A280, colorimetric assays (Bradford, 1972; Smith et al., 1985) |
| Potency | Functional ELISA, affinity column, HPLC, biomimetic assays |
| Process chemicals | Various; see Table 9 |
| Environmental contamination | LAL, bioburden |
| Aggregation | Size exclusion HPLC, UV spectrum, light scattering |
| Product homogeneity, product purity | RP-HPLC, IEC-HPLC, IEF, SDS-PAGE, affinity columns |
| Proper refolding of product | HPLC, UV spectrum, SDS-PAGE, CD/ORD |

because it can be used to estimate the maximum amount of product which has adsorbed to the system (Sofer and Nystrom, 1989).

Cross-reactivity of the product content assay should be carefully assessed. For example, an activity assay for an enzyme should consider any possibility of host enzymes being detected and included in the product calculation. Quantifying a protein by using an ELISA that reacts with a single epitope may over estimate the abundance if that epitope is not all associated with intact product. For example, light chains may be overproduced when a MAb is recombinantly expressed in culture (Prior et al., 1989). An assay that measures light chain will not agree with other assays for product, since a significant portion may not be intact IgG. Quantitation by HPLC may be inaccurate if host proteins comigrate and are integrated as part of the product peak (terAvest et al., 1992).

Assays for *product quality* are performed in-process in order to monitor the consistency of the batch, or to determine how to pool fractions before proceeding to the next step (Table 8). Assays for aggregation, such as size exclusion HPLC or simple OD-based assays, are useful to monitor losses of product or stability problems. Ion exchange, HIC, and RP chromatographic assays can be useful to detect oxidation, deamidation, aggregation, the progress of refolding reactions, or errors in posttranslational processing (Dinner & Lorenz, 1979; Garnick et al., 1988; Jones & O'Connor, 1985; McEntire, 1994). Such assays may be performed in order to

**TABLE 9**  Typical Process Chemicals Encountered in Purification Processes

| Type of chemical | Examples | Risks |
|---|---|---|
| Chaotropes | 6M GCl, 8M urea, formic acid | Carbamylation (urea) |
| Organic solvents | 10–100% propanol, acetonitrile, isopropanol | Residuals in product |
| Salts | 0–3 $M$ NaCl, KCl, ammonium sulfate, Na sulfate, many others | May catalyze deamidation, aggregation |
| Cleaning solutions | 0.5 $M$ NaOH, pH 1.5–2.0 acid solutions, detergents | Incomplete removal (esp. detergents) |
| Buffering agents | Phosphate, citrate, acetate, imidazole, amino acids | Pharmacological activity |
| Antioxidants | Such as 1–10 mM EDTA | |
| Metals | Copper, zinc, nickel; from IMAC, or added to promote S–S bonding in refolding operation | Oxidative damage to Met, Cys, other residues |
| Refolding agents | Such as glutathione, DTT, Na tetrathionate, beta-mercapto-ethanol, cysteine | Incomplete removal; covalent modification of protein |
| Media components | Such as serum, transferrin, anti-foam, albumin, growth factors, protein hydrolyzates, insulin, methotrexate | Biological risks (adventitious agents); pharmacological activities, toxicity |

choose fractions to pool from a separation, to determine when to harvest a reactor, or to verify or validate process performance. They may be performed in a qualitative or quantitative mode.

Many of these assays will be similar to assays performed on final or bulk product. However, they are more complex to validate because of the need to quantify product in the presence of contaminants and in a variety of different buffers that may include extremes of pH, ionic strength, or the presence of process chemicals (see Table 9).

## B.  In-Process Assays for Contaminants

*Contaminant assays* are also important in-process assays, and often yield the most information when performed in-process instead of on final product. For example, testing for DNA in final product may shown no detectable

TABLE 10  Examples of Contaminants Assays That May Be Performed In-Process

| Analyte | In-process methods | Comments |
|---|---|---|
| Metals | Colorimetric | From cell culture media, raw materials, or IMAC chromatography. Atomic absorption is performed on final product |
| DNA | Hybridization (specific to host cell), threshold (not specific) | Host cell derived contaminant. Direct assay and spike challenge experiments |
| Host proteins | Immunoassays (Western, ELISA, ILA) | Host cell by-products which may be immunogenic, toxic |
| Methotrexate, albumin, transferrin, insulin | Immunoassays or HPLC | Examples of possible cell culture media derived process chemicals |

DNA, yet be unable to prove that DNA has been reduced to less than 100 pg, or 10 ng, per dose (the current FDA and WHO standards). On the other hand, an early process intermediate may contain a detectable level which can be compared to validation clearance data to demonstrate that each batch will meet a standard which cannot be measured into the final product (King and Panfili, 1991; Briggs & Panfili, 1991; Levine et al., 1992). The in-process testing permits a meaningful calculation of the actual level to expect in each batch of final product, while testing the final product verifies that DNA cannot be detected. Similarly, host cell antigen assays can be used to validate the performance of a purification step when direct techniques such as gels and HPLC methods cannot quantify contaminants (Anicetti et al., 1986).

Some contaminant assays need to be performed on each batch to decide whether the product is ready for the next step. For example, it may be important to product quality to verify that an oxidizing metal has been completely removed before the product is exchanged into a buffer that may not protect it from oxidation. More commonly, an in-process assay for protein concentration will be performed after an ultrafiltration step used to concentrate product, and repeated as needed until the target concentration is achieved. The assay must be validated to a precision and accuracy suitable to the target ranges required by the process. These standards may be quite different than those required of final product release assays.

## C. Specifications Development

### 1. Bulk and Final Products

Bulk and final product are characterized and released with a battery of high resolution techniques, which are described in Chapter 6. The resulting profile of the product will demonstrate that the product is suitable in purity and homogeneity, and that toxic contaminants have been removed to safe levels (Federici, 1994). When chromatographic methods are used to analyze bulk or final product, they may resemble or differ from chromatography used in the process. In general, the most information is obtained when the analytical method used to determine purity is orthogonal to the purification method, as for example when RP-HPLC is used to analyze a product purified using ion exchange, affinity, or hydrophobic interaction chromatography. However, since process columns are loaded with much more protein per volume, their resolution is lower, and an identical analytical method may be useful to verify that pooling and chromatography resulted in product which met its purity criteria by that method.

Bulk and final products will have strict specifications set on their analytical profiles by the time of licensure. These specifications are primarily set by requirements that the product be safe, active, consistent, and fit its therapeutic regimen. Another set of criteria are determined by the process which produced the product, which has already set the upper limits on purity, yield, and homogeneity. An ideal process provides product which is *more* consistent than the patient actually needs (see Figure 5). The difference provides a safety factor which allows for assay variation, stability changes in the product, and for the unknown events which could occur during shipping, warehousing, and reconstitution of the product.

### 2. Linking In-Process Monitoring and Final Specifications

Ultimately, process performance will be monitored and action and alert limits may be set on several process intermediates. In-process action limits are set wide initially and then tightened up as more data on the process become available. Minimum *in-process specifications* are set to meet the safety requirements of the patient. Tighter process limits may be set for economic and other reasons where the process is known to be sensitive to

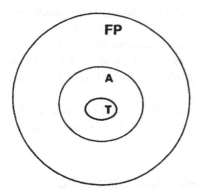

FIGURE 5  Illustration of the ideal relationship between final product specification (FP), allowed range of process settings (A), and process target setting (T). The process is aimed to run at T, but is permitted to run within A without action being taken. This guarantees that the final product will fall well within its release specifications and helps assure that most batches will pass final testing.

a particular parameter, or where data collected on multiple batches indicates that the process consistently performs within a narrower range. In a manufacturing plant, process action limits are based on concepts of statistical process control, and these may be tighter than either final product specifications or safety requirements. Exceeding these action limits does not necessarily fail a batch, yet it may indicate a need for equipment calibration, repair, personnel training, or other adjustments to the process.

### D. Using Assays to Support Process Changes

As the process is run repeatedly during development and manufacturing, the data usually suggest ways to improve the process. The act of scaling up or changing equipment may require that changes be introduced. When process changes are proposed, the process must be performed with the new changes and multiple assays performed both in-process and on final product in order to prove that the modified process equals or exceeds the known process in yield, purity, potency, and stability of the final product. The quality of a process intermediate, in contrast, may increase or decrease due to a set of process changes, provided the change is consistent and well understood. Because biological activity assays often have high variability, it may be difficult to measure a change in potency; consequently, many biochemical assays are performed to ensure that no detectable biochemical changes have resulted. Stability studies may be the most sensitive means of evaluating whether a process modification has altered the product. Changes in stability are often not predictable; some products may be less stable when at a higher purity.

An evaluation of a proposed process modification thus involves the process scientists and engineers working closely with scientists who perform and evaluate stability studies.

## VIII.  VALIDATION, PROCESS CONTROL, AND ECONOMICS

### A.  Economic Considerations in Devising a Purification Process

Cost considerations should be included in the evaluation of purification steps of a process early during development (Sofer and Nystrom, 1989; Rudge & Ladisch, 1986; Jungbauer, 1993). It is important to optimize cost for the final manufacturing scale, rather than for the scale being used during development and early clinical production. Since the dose of the drug is not known early, it may not be possible to clearly state the cost goals for a process until very late in development.

The process needs to be designed to be robust and to be practical when transferred to different facility or equipment (Levine et al., 1992; Jungbauer, 1993). Robustness may be defined in terms of the success rate of multiple process runs and also in terms of the degree of manual intervention required to control the process. Yield improvements may have to give way before the ability to tolerate fluctuations in the process parameters (Figure 4). The alternative may be expensive control equipment, which is a capital investment. The market for the product will often determine whether such control equipment is cost-effective (A. Lubiniecki, P. Shadle, J. Erickson, C. Nolan, unpublished). For biotechnology products, the data are beginning to emerge.

Relative costs have changed greatly as technology has improved. In mammalian cell culture expression systems, cell culture costs, which used to include very high media costs for serum-containing media, now have been dramatically reduced for many products. At the same time, the expression level has been pushed to higher levels, reaching 1 g/L in mammalian cell cultures (Robinson et al., 1993). These changes have had a significant impact on the cost of production. Similarly, *E. coli*, yeast, and insect cell expression systems have improved in cost due to improvements in control of expressioin, the use of defined media, and the use of fed-batch cultures to increase productivity in bioreactors.

Purification costs have changed in less well-defined ways, because proteins and their recovery processes vary greatly. Improvements in efficiency and speed brought about by chemically resistant, fast-flowing chromatography resins have improved throughput and ease of sanitization, and have reduced the volumes of cleaning solution and other buffers needed (Jungbauer, 1993). However, these changes alone do not have a large cost impact because the main cost of a biological product is the fixed capital cost (Yang et al., 1993; Erickson, 1993). The ability to use resins for hundreds of cycles, compared to the 10 to 20 cycles often used in the past, reduces raw materials costs and makes apparently high-cost affinity resins affordable in manufacturing environments. Increases in column capacity and yield likewise affect the cost of goods. But the main cost savings usually come from devising a process that has fewer steps in it, is robust enough to have a low failure rate, and makes efficient use of labor and capital.

As the scale of production increases, the relative costs of supplies and labor usually decrease. Thus, cost reductions will yield different dividends depending on the scale of production. Some changes may not be significant during development or early market launch, but could be significant after a tenfold or larger scale up in production. All cost estimations, therefore, should be designed to evaluate current and planned scales of production.

An interesting issue to consider is to compare a process with one affinity column to a process for the same product with two nonaffinity steps. Which process is more costly to run? The answer depends on several factors. In general, the capital costs dominate, suggesting that the affinity-based process is probably less costly to run, since it has fewer and smaller tanks, pumps, and chromatography controllers (Erickson, J.C., personal communication; Jungbauer, 1993). The likely higher cost of chromatography resin can be minimized by validating its use for a large number of cycles. When the lifetime is a few hundred cycles, the cost of resin per gram of product is insignificant compared to capital and overhead.

It is important to also consider QC costs, since these are significant and may well be higher for an affinity step than for a nonaffinity process step such as ion exchange, which it might replace. Added QC or validation costs may include running assays for leaching affinity ligand, defining and validating cleaning and storage conditions, and investigating any safety issues brought into the process by, for example, an animal-derived protein or a metal used as an affinity ligand. In the worst case, added cost might include a second purification step added solely to remove agents leaching from the affinity column, and/or the manufacturing cost of purifying the affinity ligand.

## B.  Process Validation and Control

As a purification process is developed, validation efforts begin. During process validation, a series of activities are performed that demonstrate that (1) the process when operated within defined ranges of critical parameters will consistently give product that is acceptable; (2) process variation has been estimated by performing replicate runs at full scale; and (3) challenge or spike studies at small scale have demonstrated confidence in the capability of the process to remove adventitious agents, host contaminants, process chemicals, and other possibly toxic agents to acceptably low levels (Sofer and Nystrom, 1989; Levine et al., 1992; Jones and O'Connor, 1985).

The development effort should identify those operating parameters that are critical to product quality and yield, develop in-process and final product assays that are relevant to the goals of each step of the process, and determine the general sensitivities of the process. This information then is used to design the validation experiments and to set action and alert limits on process ranges (Chapman, 1984). These limits are designed to (1) increase confidence that the final product will meet its release specifications; and (2) launch investigations when the process is performing significantly different than expected, even if well within a safe range. Figure 5 shows a representation of the theoretical relation between the process target settings, in-process limits, and final product attributes.

*Challenge studies* are performed for key contaminants, where it is important to demonstrate a high level of confidence that risk is insignificant. For a contaminant such as DNA, where no more than 100 ng per dose is acceptable, existing assays may not detect to this level for a product that is given in high doses. Therefore, a high concentration of host DNA is spiked into a process intermediate, and a scaled-down purification unit operation is used to demonstrate that the process has excess clearance capacity compared to the amount of DNA usually found in the unprocessed bulk (Levine et al., 1992). Clearances determined in this way on individual unit operations are combined to calculate the total clearance of the process mathematically. Such process validation experiments may greatly reduce the QC testing burden in manufacturing, where a process is shown to have a high excess capacity and to have consistency in its clearance of the significant contaminants. The clearance data are more powerful than a "not detectable" result obtained on final product, since they indicate the safety factor.

*Process control* is demonstrated by showing that the critical parameters can be effectively controlled by the equipment and procedures used, and that the outcome (purity, yield, potency, and stability) is acceptable and its variability is constant. The need for process control may lead to setting action limits that are tighter than the actual requirements of the process, but reflect the fact that the equipment can and does perform within that range. For example, a buffer LAL action limit may be set at the 1 to 5 EU/mL level based on process monitoring information, even though the product at that step is a bacterial lysate containing $10^6$ EU/mL. While the process does not require this level of control to assure that the purified product will meets its endotoxin specifications, a buffer level of ten times the usual level would indicate a problem with some system that should be investigated.

## C. Effects of Scaling Up and Process Changes

As a product progresses through clinical trials, the process scientist develops changes that may improve the process yield, the robustness of the process, and the efficiency of the process. Steps may be added to improve the removal of contaminants, or the time taken to purify the product may increase or decrease. The equipment may change in design because of scale up to provide larger quantities. Typical equipment changes may expose the product to stainless steel instead of glass and to new types of filters and tubing having the potential to leach previously unknown materials during cleaning or use. These process changes, taken together, impact on the design of similar improvements to the formulation of the final product that the formulation scientist may be making at the same time. Changes to product stability may be more difficult to predict or assess than changes in yield, purity, or potency. While major process changes clearly trigger a full investigation of stability, minor changes may be overlooked. For example, qualifying a holding step on a process intermediate may affect the eventual stability of final product that was held and then purified. Minor changes to column length that may be made when scale up dimensions cannot be maintained in existing equipment may increase or decrease purity, or may have no measurable effect. In the absence of a measurable effect on purity, is a full stability study required or not? The answer depends to a large degree on the nature of the column step, and the other data available about the product and the purification process. A matrix approach may be essential to evaluate the important changes without assaying thousands of samples.

## D. Summary and Conclusions

The isolation and purification process for a recombinant protein or peptide is designed to exploit the differences in physical and chemical properties of the product protein and the contaminants and impurities from which it must be separated. A recovery process will typically consist of a number of linked unit operations, which together reduce the volume, remove protein, nucleic acid, and other contaminants, and remove or reduce the levels of product-related impurities that were generated during synthesis in the cell, fermentation, and purification.

### 1. Process History of a Protein Affects Its Stability and Purity

At the end of the recovery process, the bulk product is highly purified and in a buffer which is amenable to storage under defined conditions until formulation and secondary manufacturing are performed. The protein has been exposed to a number of different conditions of pH, buffer composition, and exposure to contaminating proteases. The protein may have been refolded to an active conformation. Considerable microheterogeneity may exist, especially in a large, complex glycoprotein; some of it resembles the endogenous molecule in vivo, while some may have been caused by novel chemical conditions encountered only in vitro. Either type of microheterogeneity may affect the potency, pharmacokinetics, stability, or immunogenicity of the product. It is usually not possible to characterize each product variant for all of these properties, and so the safety of the product rests on the ability of the process to consistently produce a consistent composition.

The safety of the product has also been affected by its purification history. Process chemicals have been used, which might have changed the protein's conformation, chemical integrity, or biological potency. Residuals of these chemicals may be present in the bulk product, and may affect its ability to be formulated, lyophilized, or stored.

### 2. Process Development Generates Useful Data

Much of the data which were collected during process development are very useful to the formulation scientist. In-process assays and purification methods which are used at full scale or were used early in development may be adapted as product release assays or stability-indicating methods. Many pathways of degradation of the protein product, and chemical conditions under which degradation is accelerated, have been identified and partially defined. Reagents such as process side fractions enriched for host cell proteins and other contaminants or impurities have been generated, and may be available for study. In addition, there may be atypical batches in which unplanned deviations occurred, which could contain important information on the consequences of exceeding normal process ranges for pH, time, or other parameters.

## 3. Process Validation Aids in Interpretation of Stability Data

As the process is scaled up, transferred to a bulk manufacturing plant, and validated for a license application, additional data on the process are generated. Process validation establishes which parameters are critical to the outcome of the process and defines ranges within which a consistent, acceptable product can be made. Replicate runs at full scale demonstrate that the process is in a state of control and is capable of making product which meets its release specs. for purity, potency, and stability. These activities generate large numbers of stability samples, both of product and of process intermediates. Interpreting the results involves both the groups performing stability studies and those familiar with the process. These groups also start working together more closely to integrate the license application package. In particular, the rationale for setting in-process action and alert limits should be related to the process, and to the release testing of the product.

As more experience is gained with the process, validation experiments can be done which relieve the testing burden. For example, testing for a process chemical may be dropped once process validation demonstrates an excess clearance capacity and replicate lots of bulk or final product test negative for the chemical.

In conclusion, the process of expressing and purifying the product defines the consistency, purity, and homogeneity of the product. Setting up relevant assays to evaluate the product and to understand the inherent variability in the product is the challenge for the formulation scientist. This chapter has given an overview of the approaches and methods used to develop a purification process, and has described the utility of process thinking and process information.

## ACKNOWLEDGMENTS

The author would like to acknowledge Ronald Wetzel and Marcia Federici for useful discussion and reading the manuscript, and Anthony Lubiniecki and Sally Brewster for their support and inspiration.

# REFERENCES

Abour RA, Aunins JG, Buckland BC, Dephillips PA, Hagen AJ, Hennessey JP, Junker B, Lewis JA, Oliver CN, Orella CJ, Sitrin RD, Newell O. Production of hepatitis A virus vaccine—by large scale culturing of virus followed by purification to obtain highly-pure HAV protein. Eur Pat 583142, Merck & Co., 1994.

Anicetti VR, Fehskens EF, Reed BR, Chen AB, Moore P, Geier MD, Jones AJS. Immunoassay for the detection of *E. coli* proteins in recombinant DNA derived human growth hormone. J Immunol Methods 1986; 91:213–224.

Arathoon WR, Builder SE, Lubiniecki AS, van Reis RD. U.S. Pat 90-522073, 1990.

Arnold FH. Metal-affinity separations: A new dimension in protein processing. Bio-Technology 1991; 9:151–156.

Babbitt PC, West BL, Buechter DD, Kuntz ID, Kenyon GL. Bio/tech 1990; 8:945–949.

Bennett AD, Rhind SK, Lowe PA, Hentschel CG. Eur Pat Appl 0131363, 1984.

Benson BJ, Buckley D, Lesikar D, Naidu A, Silverness KB. US Pat 5,258,496, assigned to Scios Nova Inc., 1993.

Bishop CA, Harding DRK, Meyer LJ, Hancock WS, Hearn MTW. High-performance liquid chromatography of amino acids, peptides and proteins. XXI. The application of preparative reversed-phase high-performance liquid chromatography for the purification of a synthetic underivatised peptide. J Chrom 192:222–227, 1980.

Bogosian G, Violand BN, Dorward-King EJ, Workman WE, Jung PE, Kane JF. Biosynthesis and incorporation into protein of norleucine by *Escherichia coli*. J Biol Chem 264:531–539, 1989.

Bornstein P, Balian G. Cleavage at Asn-Gly bonds with hydroxylamine. Methods in Enzymology 14:132–145, 1976.

Bradford MM. A rapid and sensitive method for the quantitation of microgram quantities of protein utilizing the principle of protein-dye binding. Analytical Biochem 72:248–254, 1976.

Briggs J, Panfili P. Quantitation of DNA and protein impurities in biopharmaceuticals. Anal Chem 63:850–859, 1991.

Builder S. Recovery of Biological Products VII. Engineering Foundation Symposium, San Diego, CA, Sept. 1994, paper #9.4.

Builder SE, van Reis R, Paoni NF, Ogez JR. Process development and regulatory approval of tissue-type plasminogen activator. In: Durand G, Bobichon L, Florent J, eds. 8th Int Biotechnol Symp. Soc Fran Microbiol, Paris, France, 1988, 1:644–659.

Canova-Davis E, Eng M, Mukku V, Reifsnyder DH, Olson CV, Ling VT. Chemical heterogeneity as a result of hydroxylamine cleavage of a fusion protein of human insulin-like growth factor-1. Biochem J 285:207–213, 1992.

Carter P. Site-specific proteolysis of fusion proteins. In: Ladisch MR, Willson RC, Painton CC, Builder SE, eds. American Chemical Society Symposium Series No. 427. ACS, Washington DC, 1990, pp 181–193.

Chadha KC, Grob PM, Mikulski AJ, Davis LR, Sulkowski E. Copper chelate affinity chromatography of human fibroblast and leucocyte interferons. J Gen Virol 43:701–706, 1979.

Chance RE, Kroeff EP, Hoffmann JA. Chemical, physical, and biological properties of recombinant human insulin. In: Gueriguian JL, ed. Insulins, Growth Hormone, and Recombinant DNA Technology. New York: Raven Press, 1981, pp 71–86.

Chance RE, Hoffmann JA, Kroeff EP, Johnson MG, Schirmer EW, Bromer WW, Ross MJ, Wetzel R. The production of human insulin using recombinant DNA technology and a new chain combination procedure. Proceedings of 7th American Peptide Symposium, Rockford, Ill, 1981, pp 721–728.

Chapman KG. The PAR approach to process validation. Pharmaceutical Technology 8:22–36, 1984.

Chase HA, Draeger NM. Affinity purification of proteins using expanded beds. J Chromatogr 597:129–145, 1992.

Cleland JL, Builder SE, Swartz JR, Winkler M, Chang JY, Wang DIC. Polyethylene glycol enhanced protein refolding. Bio/Technol 10:1013–1019, 1992.

Dahlman K, Strömstedt P-E, Rae C, Jörnvall H, Flock J-I, Caarlstedt-Duke J, Gustafsson J-Å. High level of expression in *E. coli* of the DNA-binding domain of the glucocorticoid receptor in a functional form utilizing domain-specific cleavage of a fusion protein. J Biol Chem 264:804–809, 1989.

Dean PDG, Johnson WS, Middle FA, eds. Affinity chromatography, a practical approach. New York: IRL Press (imprint of Oxford University Press, Inc.), 1985.

Dinner A, Lorenz L. High performance liquid chromatographic determination of bovine insulin. Analytical Chemistry 561:1872–1873, 1979.

Duffy SA, Moellering BJ, Prior GM, Doyle KR, Prior CP. Recovery of therapeutic-grade antibodies: Protein A and ion-exchange chromatography. BioPharm 8:35–47, 1989.

Erickson JC. Economic analysis of biopharmaceutical processes. ACS Annual Symposium, San Diego, California, 1993.

Erickson JC, Finch JD, Greene DC. Direct capture of recombinant proteins from animal cell culture using a fluidized bed absorber. In: Griffiths B, Spier RE, Berthold W, eds. Animal cell technology—products for today, prospects for tomorrow. Oxford, UK: Butterworth & Heinemann, 1994, pp 557–562.

Ezzedine M, Lawny F, Vijayalakshmi MA. Purification and depyrogenation of anti-hemophilic FVIII:C from human plasma. In: Rivat C, Stoltz J-F, eds. Biotechnol of Blood Proteins. Colloq. INSERM, Paris, France, 1993, 227: pp 115–123.

Federici MM. The quality control of biotechnology products. Biologicals 22:51–159, 1994.

Federici MM, Garnick RL. A perspective on reference standard and reference material requirements in biotechnology-derived pharmaceutical protein products. Pharmacopeial Forum 17:2683–2687, 1991.

Fontana A, Toniolo C. The chemistry of tryptophan in peptides and proteins. In: Herz W, Grisebach H, Kirby GM, eds. Progress in the Chemistry of Organic Natural Products, Vol. 33. New York: Springer-Verlag, 1976, pp. 309–449.

Forsberg G, Holmgren E, Bergdahl K, Brobjer M, Persson P, Gautvik K, Hartmanis M. Thrombin and H64A subtilisin cleavage of a fusion protein for production of human recombinant parathyroid hormone. J Prot Chem 10:517–526, 1991.

Forsberg G, Baastrup B, Rondahl H, Holmgren E, Pohl G, Hartmanis M, Lake M. An evaluation of different enzymatic cleavage methods for recombinant fusion proteins, applied on des(1-3)insulin-like growth factor I. J Prot Chem 11: 201–211, 1992.

Garnick RL, Solli NJ, Papa PA. The role of quality control in biotechnology: An analytical perspective. Analyt Chem 60:2546–2557, 1988.

Gilligan JP, Warren TG, Koehn JA, Young SD, Bertelsen AH, Jones BN. Purification of a fusion protein containing recombinant human calcitonin. BioChromatography 2:20–27, 1987.

Glaser CB, Karic L, Huffaker T, Chang R, Martin J. Studies on the disulfide region of $\alpha_1$-protease inhibitor. Int J Peptide Protein Res 29:56–62, 1982.

Goeddel DV, Kleid DG, Bolivar F, Heyneker HL, Yansura DG, Crea R, Hirose T, Kraszewski A, Itakura K, Riggs AD. Expression in E. coli of chemically synthesized genes for human insulin. Proc Natl Acad Sci USA 76:106–110, 1979.

Goffe RA, Zale SE, O'Connor JL, Kessler SB. Cohen CM. PCT Int Appl US88-265061 881031, 1988.

Gracy RW. Effects of aging on proteins. In: Ahern TJ, Manning MC, eds. Stability of Protein Pharmaceuticals. Part B: In Vivo Pathways of Degradation and Strategies for Protein Stabilization. New York: Plenum Press, 1992, pp. 119–145.

Halenbeck R, Shadle PJ, Lee P-J, Lee M, Koths K. Purification and characterization of recombinant human macrophage colony-stimulating factor and generation of a neutralizing antibody useful for Western analysis. J Biotechnol 8:45–58, 1988.

Hanna LS, Pine P, Reuzinsky G, Nigam S, Omstead DR. Removing specific cell culture contaminants in a MAb purification process. BioPharm 10:33–37, 1991.

Hilaireau P, Cohn H, Cox GB. Industrial continuous liquid chromatography. Indust Chrom News (ProChrom) 5:17–24, 1990.

Hochuli E. Large-scale chromatography of recombinant proteins. J Chromatogr 444:293–302, 1988.

Hochuli E, Bannwarth W, Dobeli H, Gentz R, Stuber D. Genetic approach to facilitate purification of recombinant proteins with a novel metal chelate absorbent. Bio/Technology 6:1321–1325, 1988.

Hoess A, Arthur AK, Wanner G, Fanning E. Recovery of soluble, biologically active recombinant proteins from total bacterial lysates using ion exchange resin. Bio/Technology 6:1214–1217, 1988.

Holmberg L, Bladh B, Astedt B. Purification of urokinase by affinity chromatography. Biochem Biophys Acta 445:215–223, 1976.

Hood W, DeLa Morena E, Grisolia S. Increased susceptibility of carbamylated glutamate dehydrogenase to proteolysis. Acta Biol Med Germ 36:1667–1672, 1977.

Hrinda ME, Tarr C, Curry W, Newman J, Schreiber AB, D'Alisa R. Monoclonal antibody affinity purification of plasma proteins removes viral contaminants. In: Stoltz JF, Rivat C, eds. Biotechnology of plasma proteins. Colloq IN-SERM, Paris, France, 1989, 175: pp 413–418.

Hsieh P, Rosner MR, Robbins PW. Host dependent variation of asparagine-linked oligosaccharides at individual glycosylation sites of Sindbis virus glycoproteins. J Biol Chem 258:2548–2554, 1983.

Johnson BA, Aswad DW. Fragmentation of isoaspartyl peptides and proteins by carboxypeptidase Y. Biochemistry 29:4373–4380, 1990.

Johnson BA, Shirokawa JM, Hancock WS, Spellman MW, Basa LJ, Aswad DW. Formation of isoaspartate at two distinct sites during in vitro aging of human growth hormone. J Biol Chem 264:14262–14271, 1989.

Johnson PR, Stern NJ, Eitzman PD, Rasmussen JK, Milbrath DS, Gleason RM, Hogancamp RE. Reproducibility of physical characteristics, protein immobilization and chromatographic performance of 3M Emphaze biosupport medium AB1. J Chrom 667:1–9, 1994.

Jones AJS, O'Connor JV. Control of recombinant DNA produced pharmaceuticals by a combination of process validation and final product specifications. Develop biol Standard 59:175–180, 1985.

Jungbauer A. Preparative chromatography of biomolecules. J Chromatography 639: 3–16, 1993.

Kamihira M, Kaul R, Mattiasson B. Purification of recombinant protein A by aqueous two-phase extraction integrated with affinity precipitation. Biotechnol and Bioengin 40:1381–1387, 1992.

Kane JF, Hartley DL. Formation of recombinant protein inclusion bodies in *Escherichia coli*. Tibtech 6:95–101, 1988.

Kemmler W, Peterson JD, Steiner DF. Studies on the conversion of proinsulin to insulin. I. Conversion in vitro with trypsin and carboxypeptidase B. J Biol Chem 246:6786–6791, 1971.

Keshavarz E, Hoare M, Dunnill M. Biochemical engineering aspects of cell disruption. In: Verrall MS, Hudson MJ, eds. Separation for Biotechnology. Chichester, UK: Ellis Horwood, 1987, pp. 62–79.

King RS, Panfili PR. Influence of fragment size on DNA quantitation using DNA-binding proteins and a sensor-based analytical system: applications in the testing of biological products. J of Biochem and Biophysical Methods 23: 83–93, 1991.

Knight E, Fahey D. Human fibroblast interferon—an improved purification. Biol Chem 256:3609–3613, 1981.

Kroeff EP, Owens RA, Campbell EL, Johnson RD, Marks HI. Production scale purification of biosynthetic human insulin by reversed phase high performance liquid chromatography. J Chrom 461:45–61, 1989.

Kunitani M, Johnson D, Snyder LR. Model of protein conformation in the reversed-phase separation of interleukin-2 muteins. J Chrom 371:313–333, 1986a.

Kunitani M, Hirtzer P, Johnson D, Halenbeck R, Boosman A, Koths K. Reversed-phase chromatography of interleukin-2 muteins. J Chromatography 359: 391–402, 1986b.

Lai P-H, Strickland TW. U.S. Patent No. 4,667,016, 1987.

Lee JM, Ullrich A. Eur. Pat. Appl. 0128733, 1984.

Levine HL, Tarnowski SJ, Dosmar M, Fenton DM, Gabler R, Gardner JN, Hageman TC, Lu P, Sofer G, Steininger B. Industry perspective on the validation of column-based separation processes for the purification of proteins. J of Parenteral Science and Technology Suppl 46:87–97, 1992.

London J, Skrzynia C, Goldberg ME. Renaturation of *Escherichia coli* tryptophanase after exposure to 8 M urea. Evidence for the existence of nucleation centers. Eur J Biochem 47:409–415, 1974.

Lu HS, Tsai LB, Kenney WC, Lai P-H. Identification of unusual replacement of methionine by norleucine in recombinant interleukin-2 produced by *E. coli*. BBRC 156:807–813, 1988.

Lubiniecki AS, ed. Large-Scale Mammalian Cell Culture Technology. New York: Marcel Dekker, 1990.

Lubiniecki AS, Wiebe ME, Builder SE. Process validation for cell culture derived pharmaceutical proteins. In: Lubiniecki AS, ed. Large-Scale Mammalian Cell Culture Technology. New York: Marcel Dekker Inc, 1990, pp. 515–541.

Lubiniecki AS, Dinowitz M, Nelson E, Wiebe M, May L, Ogez J, Builder S. Endogenous retroviruses of continuous cell substrates. Symp on Continuous Cell Lines as Substrates for Biologicals. Develop Biol Standard 70:187–191, 1988.

Maiorella BL, Ferris R, Thomson J, White C, Brannon M, Hora M, Henriksson T, Triglia R, Kunitani M, Kresin L, Dollinger G, Jones R, Senyk G, Young J, Moyer B, McEntire J, Dougherty J, Monica T, Goochee C. Evaluation of product equivalence during process optimization for manufacture of a human IgM monoclonal antibody. Biologicals 21:197–205, 1993.

Maisano F, Testpro SA, Grandi G. Immobilized metal-ion affinity chromatography of human growth hormone. J Chromatogr 472:422–427, 1989.

Mannweiler K, Titchener-Hooker NJ, Hoare M. Biochemical engineering improvements in the centrifugal recovery of biological particles. IChemE Symp: Advances in biochemical engineering, Newcastle, UK, 1989, pp 105–117.

Mariani M, Tarditi L. Validating the preparation of clinical monoclonal antibodies. Bio/Technology 10:394–396, 1992.

Marston FAO. The purification of eukaryotic polypeptides synthesized in *Escherichia coli*. Biochem J 240:1–12, 1986.

Martin K. Benzon-nuclease—characteristics and application of the new type of endonuclease for the elimination of disturbing nucleic acids. Schweiz Lab-Z 48:199–201, 1991.

McEntire J. Biotechnology product validation, part 5: Selection and validation of analytical techniques. BioPharm 7:68–80, 1994.

Means GE, Feeney RE. Chemical Modification of Proteins. San Francisco: Holder-Day, 1971.

Melander WR, Corradini D, Horvath C. Salt-mediated retention of proteins in hydrophobic interaction chromatography. Application of solvophobic theory. J Chromatog 317:67–85, 1984.

Morehead H, Johnston PD, Wetzel R. Roles of the 29–138 disulfide bond of subtype A of human α interferon in its antiviral activity and conformational stability. Biochemistry 23:25–35, 1984.

Nachman M, Azad ARM, Bailon P. Efficient recovery of recombinant proteins using membrane-based immunoaffinity chromatography (MIC). Biotechnol and Bioengin 40:564–571, 1992.

Nagai K, Perutz MF, Poyart C. Oxygen binding properties of human mutant hemoglobins synthesized in *Escherichia coli*. Proc Natl Acad Sci USA 82:7252–7255, 1985.

Narayanan SR, Crane LJ. Affinity chromatography supports: a look at performance requirements. Tibtech 8:12–16, 1990.

Ogez JR, Nodgdon JC, Beal MP, Builder SE. Downstream processing of proteins: Recent advances. Biotechnol Adv 7:467–488, 1989.

Parekh RB, Tse AGD, Dwek RA, Silliams AF, Rademacher TW. Tissue-specific N-glycosylation, site-specific oligosaccharide patterns and lentil lectin recognition of rat Thy-1. EMBO J 6:1233–1244, 1987.

Parekh RB, Dwek RA, Thomas JR, Opdenakker G, Rademacher TW. Cell-type-specific and site-specific N-glycosylation of type I and type II human tissue plasminogen activator. Biochemistry 28:7644–7662, 1989a.

Parekh RB, Dwek RA, Thomas JR, Opdenakker G, Rademacher TW. N-glycosylation and in vitro enzymatic activity of human recombinant tissue plasminogen activator expressed in Chinese hamster ovary cells and a murine cell line. Biochemistry 28:7670–7678, 1989b.

Points to Consider in the characterization of cell lines used to produce biologicals. Food and Drug Administration, May 1993.

Points to consider in the manufacture and testing of monoclonal antibody products for human use. Food and Drug Administration, 1994.

Prior CP, Doyle KR, Duffy SA, Hope JA, Moellering BJ, Prior GM, Scott RW, Tolbert WR. The recovery of highly purified biopharmaceuticals from perfusion cell culture bioreactors. J Parenter Sci Tech 43:15–23, 1989.

Quinlan GJ, Coudray C, Hubbard A, Gutteridge JM. Vanadium and copper in clinical solutions of albumin and their potential to damage protein structure. J Pharm Sciences 81:611–614, 1992.

Reisman HB. Economic analysis of fermentation processes. Biotechnologists & Microbiologists Series. Boca Raton, FL: CRC Press, 1988.

Righetti PG, Barzaghi B, Sarubbi E, Soffientini A, Cassani G. Charge heterogeneity of recombinant pro-urokinase and urinary urokinase, as revealed by isoelectric focusing in immobilized pH gradients. J Chromatogr 470: 337–350, 1989.

Rivier J, McClintock R. Reversed-phase high-performance liquid chromatography of insulins from different species. J Chrom 268:112–119, 1983.

Rivier J, McClintock R, Galyean R, Anderson H. Reversed phase HPLC: Preparative purification of synthetic peptides. J Chromatography 288:303–328, 1984.

Robinson DK, Chan CP, Ip CCY, Seamans TC, Lee DK, Lenny AB, Tung J-S, DiStefano DJ, Munshi S, Gould SL, Tsai PK, Irwin J, Mark GE, Silberklang M. Product consistency during long-term fed-batch culture. In: Spier RE, Griffiths JB, Berthold W, eds. Animal Cell Technology, Products of Today, Prospects for Tomorrow. Proceedings of ESACT 12th meeting, San Diego, California, 1993, pp 763–767.

Rudge ST, Ladisch MR. 10. Process Considerations for Scale-up of Liquid Chromatography and Electrophoresis. Separation, Recovery, and Purification in Biotechnology, ACS, 1986, pp 123–140.

Saito Y, Ishii Y, Koyama S, Tsuji K, Yamada H, Terai T, Kobayashi M, Ono T, Niwa M, Ueda I, Kikuchi H. Bacterial synthesis of recombinant alpha-human atrial natriuretic polypeptide. J Biochem 102:111–122, 1987.

Schoner RG, Ellis LF, Schoner BE. Isolation and purification of protein granules from *E. coli* cells over-producing bovine growth hormone. Biotechnology 3: 151–154, 1985.

Schwinn H, Smith A, Wolter D. Progress in purification of virus-inactivated factor VIII concentrates. Arzneim.-Foreschung 39:1302–1305, 1989.

Scopes RK. Protein Purification, Principles and Practice, 2nd ed. New York: Springer-Verlag, 1982, pp 1–66.

Seely RJ, Young MD. Large-scale refolding of secretory leukocyte protease inhibitor. Protein Refolding. ACS Symp Ser 470:206–216, 1991.

Shadle PJ, Silverness K, Wallace L, Scheuermann T, Tarnowski SJ. Metal chelate affinity chromatography can damage recombinant proteins. Presented at: Recovery of Biological Product VI, Engineering Foundation, Interlaken, Switzerland, September 1992.

Shaked Z, Wolfe S. U.S. Patent 4 450 787, assigned to Chiron Corporation, 1985.

Sharma, B. [In this volume]

Shine J, Fiddes J, Lan NCY, Roberts JL, Baxter JD. Expression of cloned beta-endorphine gene sequences by *Escherichia coli*. Nature 285:4356–4461, 1980.

Sitrin R, DePhillips P, Dingerdissen J, Erhard K, Filan J. Preparative liquid chromatography, a strategic approach. LC-GC Mag Liq & Gas Chromatog 4:530–543, 1986.

Smith PK, Krohn RI, Hermanson GT, Mallia AK, Gartner FH, Provenzano MD, Fujimoto EK, Gorke NM, Olson BJ, Klink DC. Measurement of protein using bicinchoninic acid. Anal Biochem 150:76–85, 1985.

Sofer GK, Nystrom L-E. Process chromatography. San Deigo, CA: Academic Press, 1989.

Steindl F, Jungbauer A, Wenisch E, Himmler G, Katinger H. Isoelectric precipitation and gel chromatography for purification of monoclonal IgM. Enzyme and Microb Technol 9:361–364, 1987.

Steiner DF, Clark JL. The spontaneous reoxidation of reduced beef and rat proinsulins. Proc Natl Acad Sci USA 60:622–628, 1968.

Sulkowski E. Purification of proteins by IMAC. Trends in Biotechnology 3:1–7, 1985.

Szoka PR, Schreiber AB, Chan H, Murthy J. A general method for retrieving the components of a genetically engineered fusion protein. DNA 5:11–20, 1986.

terAvest AR, van Zoelen EJJ, Spijkers IEM, Osterhaus ADME, van Steenis G, van Kreyl CF. Purification process monitoring in monoclonal antibody preparation: Contamination with viruses, DNA and peptide growth factors. Biologicals 20:177–186, 1992.

Thompson SA, Fiddes JC. Chemical characterization of the cysteines of basic fibroblast growth factor. In: Baird A, Klagsbrun M, eds. The Fibroblast Growth Factor Family. Annals of the New York Academy of Sciences, 638:78–88, 1991.

Tsuji T, Nakagawa R, Sugimoto N, Fukuhara K-i. Characterization of disulfide bonds in recombinant proteins: reduced human interleukin-2 in inclusion bodies and its oxidative refolding. Biochem 26:3129–3134, 1987.

Ueno Y, Morihara K. Use of immobilized trypsin for semisynthesis of human insulin. Biotechno and Bioeng 33:126–128, 1989.

Uhlen M, Forsberg G, Moks T, Hartmanis M, Nilsson B. Fusion proteins in biotechnology. Curr Opin Biotechnol 3:363–369, 1992.

Urdal DL, Mochizuki D, Conlon PJ, March CJ, Remerowski ML, Eisenman J, Ramthun C, Gillis S. Lymphokine purification by reversed-phase high performance liquid chromatography. J Chrom 296:171–179, 1984.

Valax P, Georgiou G. In: Georgiou G, De Bernardez-Clark E, eds. Protein Refolding. Washington, D.C.: ACS, 1991, pp 99–109.

Valax P, Georgiou G. 10. Characterization and refolding of beta-lactamase inclusion bodies in *E. coli*. In: Georgiou G, De Bernardez–Clark ED, eds. Protein Refolding: Biocatalyst Design for Stability and Specificity. New York: ACS Symposium Series, 1993.

van Reis R, Builder S. UF as a purification method. At: Recovery of Biological Products VI, Engineering Foundation Symposium, Interlaken, Switzerland, September, 1992.

van Reis R, Leonard LC, Hsu CC, Builder SE. Industrial scale harvest of proteins from mammalian cell culture by tangential flow filtration. Biotechnol Bioeng (USA) 38:413–422, 1991.

Wetzel R. Protein aggregation in vivo. Bacterial inclusion bodies and mammalian amyloid. In: Ahern TJ, Manning MC, eds. Stability of Protein Pharmaceuticals. Part B: In Vivo Pathways of Degradation and Strategies for Protein Stabilization. New York: Plenum Press, 1992, pp 43–88.

White EM, Grun JB, Sun C-S, Sito AF. Process validation for virus removal and inactivation. Biopharm 4:34–39, 1991.

Williams DC, Van Frank RM, Muth WL, Burnett JP. Cytoplasmic inclusion bodies in *E. coli* producing biosynthetic human insulin proteins. Science 215:687–689, 1982.

Winkler ME, Blaber M, Bennett GL, Holmes W, Vehar GA. Purification and characterization of recombinant urokinase from *Escherichia coli*. Bio/Technol 3: 990–1000, 1985.

Yang R, Vellayudhan A, Ladisch DM, Ladisch MR. Liquid chromatography using cellulosic continuous stationary phases. Adv Biochem Eng/Biotechnol 49: 147–160, 1993.

# 4

# Preformulation Development of Parenteral Biopharmaceuticals

JOHN A. BONTEMPO

Biopharmaceutical Product Development, East Brunswick, New Jersey

## I. INTRODUCTION

Preformulation research studies of protein therapeutics encompass biopharmaceutical, physicochemical, and analytical investigations in support of subsequent stable formulations for preclinical, clinical, and market usage.

In this highly competitive protein therapeutics field, it is very important to obtain significant, measurable progress with preformulations studies in a timely manner. How extensive these studies are will depend on the availability of the crude, active drug substance and the intended route of administration. Most often, these studies begin with extremely small amounts of crude bulk active substance and, as more material becomes available with greater purity, more studies are initiated.

From an industrial point of view, the preformulation studies are designed to cover a wide range of properties in a short time to learn as much as possible, but not in great depth. The pharmaceutical formulation scientist is very much interested in identifying potential problems early enough to evaluate potential alternatives to stabilize future formulation(s).

As previously stated, there must be a strong interdisciplinary collaboration team to review, identify, and maximize the most productive leads toward formulation development. Preformulation studies are short in duration, two to three months, and some of these are performed under varying stress conditions which will be described later in this chapter. It is important to remember that no two proteins are alike and studies designs will vary case by case.

Prior to the onset of preformulations, the pharmaceutical team must review some very important factors which will have an impact on the preformulation and formulation development.

## A. Considerations of Domestic and International Distribution of the Product

Many global joint ventures and partnerships today in the biopharmaceutical industry dictate various pharmaceutical, clinical, and marketing strategies. The regulatory requirements and acceptance of formulation excipients, packaging components, unit dose versus multidose product, and stability conditions vary from continent to continent. Constituted and/or lyophilized dosage forms must also be considered. The development of formulation considerations should be on a worldwide acceptance basis.

## B. Points to Consider for Constituted Versus Lyophilized Formulations

Some of the key points to be considered for a constituted formulation are:

- A constituted formulation may be less stable than a lyophilized one
- Effect of agitation during manufacturing and shipping
- Interaction of the liquid with the inner wall of the glass vial and with the elastomeric closure
- Aggregation problems
- Head space within the vial
- Preservative effectiveness

Some of the key points to be considered for a lyophilized formulation are:

- Better stability than a constituted product
- Determination of an optimal lyophilization cycle
- Effects of residual moisture on the activity and stability of the product
- Ease of reconstitutability. Clinicians, nurses and trained home users, prefer reconstitutability of the product within two minutes.

- Stability of the reconstituted product
- Preservative effectiveness (if this is a multidose product)
- Cost effectiveness. Lyophilization technology is expensive along with cost of utilities

At the onset of preformulations studies, it is difficult to predict with certainty which of the two types of formulations will have a marketable advantage for an extended shelf life. At this early stage of development, there are usually very small amounts of the bulk active drug substance available. The formulator must make very efficient use of the active drug substance. Nevertheless, both formulations should be considered and started at the same time. Stability results should be the deciding factor as to which form will be selected for further development.

## C. Unit Dose or Multidose

The decision to select unit dose versus multidose should be based upon input from clinical investigators, focus groups, marketing surveys, and competitors' products. A multidose formulation will require significantly more time for development.

The multidose will require the screening and incorporation of compatible preservative(s) with the protein formulation. This formulation will be tested to determine if it is efficacious enough to meet the United States Pharmacopeia (USP) requirements. Meeting these requirements, it can qualify as a "multidose" for the U.S. market. However, if the formulation is also designated for international market, there are three additional factors that must be taken into account. The *first* is that for the "antimicrobial effectiveness test," a particular country may or may not accept the preservative selected. *Secondly*, the concentrations of the preservative present in the formulation may be different from the USP requirements. *Thirdly*, the time periods required for the inhibition of the bacteria and fungi strains tested may also differ. Consequently, I strongly suggest that the international regulatory requirements for compliance should be well researched and understood by the scientific and management staff. Other excipients should also be thoroughly reviewed for international acceptance.

## D. Physicochemical Factors to Be Considered for Protein Drug Formulations

Some of the most important physicochemical properties of protein drugs required for the development of parenteral preformulations and formulations are found in Table 1.

TABLE 1  Physicochemical Factors to Be Considered for Protein Drug Formulations

| Structure of the protein drug | Agents affecting stability |
| --- | --- |
| Isoelectric point | pH |
| Molecular weight | Temperature |
| | Light |
| Amino acid composition | Oxygen |
| | Metal ions |
| Disulfide bonds | Freeze–thaw |
| | Mechanical stress |
| Spectral properties | |
| Agents affecting solubility: | Polymorphism |
| Detergent | Stereoisomers |
| Salts | Filtration media compatibility |
| Metal ions | Shear |
| pH | Surface denaturation |

Since this may be an early stage of process development, some of the properties listed in Table 1 may not be available initially, simply because there was not enough time or personnel to perform the work.

## II. INITIAL PREFORMULATION STUDIES: PARAMETERS AND VARIABLES TO BE TESTED

The pharmaceutical formulation scientist will consider several factors in the preformulation designs. The data received from the Process/Purification section are reviewed for structure, pH and purity of the substance, preliminary bioassay, and an immune assay used in terms of semiquantitative measurements.

Other important information that may or may not be available are product solubility, preliminary stability, potential degradation routes. From personal experience, there is only minimal crude bulk active substance at this early stage.

### A. Initial Variables to Be Tested

Perhaps 10 or more initial preformulation combinations should be considered. The initial variables to be tested with various protein concentrations are the effects of buffer species, ionic strength, pH range, temperature, initial shear, surface denaturation, agitation, and aggregation. Since it has been well documented that protein solutions are unstable, some selective excipients from various classes of stabilizers should also be included in order to evaluate stability requirements. Stabilizers will be discussed later in this volume.

## B. Preliminary Analytical Development

In order to determine the initial stability results, it is necessary to have developed, or to have under development, analytical methods to measure the potency of the specific formulation under various experimental conditions. Ultimately some of these analytical methods will be needed to monitor stability to detect physical and chemical degradation. Regulatory compliance for the beginning of Phase I Clinical Studies may require at least two different methods that are "stability indicators," most often fully validated. Dr. Sharma, in Chapter 6, will cover the bioanalytical development.

**TABLE 2**   Bioanalytical Methods to Evaluate Initial Preformulation Development

| Method | Function |
| --- | --- |
| Bioassay | Measure of activity throughout shelf life of a formulation |
| Immunoassay | Purity assessment and measures concentration of a particular molecular species |
| pH | Chemical stability |
| SDS-PAGE (Reduced & nonreduced) | Separation by molecular weight, characterization of proteins and purity |
| RP-HPLC | Estimation of purity, identity, and stability of proteins. Separation and analysis of protein digests. |
| IEF | Determines the isoelectric point of the protein and detects modifications of the protein |
| SE-HPLC | Method of separating molecules according to their molecular size and purity determination |
| N-terminal sequencing | Elucidation of the C-terminus, identity |
| UV | Detection of individual component, concentration, and aggregation |
| CD (circular dichroism) in the UV region | Detects secondary and tertiary conformation and quantitates various structures |

However, some of the following bioanalytical methods listed in Table 2 can be applied to begin initial evaluation of the preformulations degradation (if any) under test.

## C. Experimental Conditions for the Initial Preformulation Studies

### Protein Concentration

Protein drugs are extremely potent; therefore, very low concentrations are required for their respective therapeutic levels. Dosage forms development need to be tested at varying ranges of activity. The respective concentrations may range from nanograms to micrograms to milligrams and the concentration will vary from protein to protein.

### pH Range

Initially, a range of pHs should be selected, for example, 3, 5, 7 and 9. Specific pH units will be determined during the formulation studies. The pH changes may have varying impacts on the solubility and stability of the formulation. pH control in pharmaceutical dosage forms is very critical (1). The proper pH selection is one of the key factors in developing a stable product.

### Buffers

The buffer(s) selection should be made from the USP physiological buffers list and should be selected based upon their optimal pH range. Some of these buffers are acetate pH 3.8–5.8, succinate pH 3.2–6.6, citrate pH 2.1–6.2, phosphate pH 6.2–8.2, and triethanolamine pH 7.0–9.0. These pH ranges will differ from protein to protein.

Buffer concentrations should be in the range of 0.01 to 0.1 molar concentration. As buffer concentration goes up, so does the pain upon injection. In selecting the proper buffer, phosphate should be the last in one's choice. Phosphate buffer reacts with calcium from the glass vial and zinc from the rubber stopper to cause glass laminates and eventually haziness of the solution during stability periods.

### Other Excipients to Be Considered

As it was stated previously, the objective of a preformulation study is to select potentially compatible excipients in order to hasten the development of stable formulations. Based upon protein chemical and physical instability, it is highly probable that some excipients may be included in the preformulation. In so doing, the designs of formulations to follow can be more specific in selecting the proper excipient(s) to control specific degradation pathways.

### Chelating Agents

The crude bulk protein drug during fermentation and purification steps has passed through and contacted surfaces such as metal, plastic, and glass. If metal ions are present in the liquid bulk active, it is highly recommended to use a chelating agent such as ethylenediamine tetraacidic acid (EDTA) to effectively bind trace metals such as copper, iron, calcium, manganese and others. A recommended dose of (EDTA) would be about 0.01 to 0.05%.

### Antioxidants

Since oxidation is one of the major factors in protein degradation, it is highly recommended, should the use of a specific antioxidant be required, to include into the preformulation an antioxidant such as ascorbic acid, sodium disulfide, monothio-glycerol, or alpha tocopherol. The role of an antioxidant is to deplete or block a specific chain reaction. Antioxidants will be the preferential target and eventually be depleted, or may block a specific chain reaction. Argon and/or nitrogen gas can also be used to flood the head space of a vial or ampule during sterile filling to prevent or retard oxidation. A recommended antioxidant dose would be about 0.05 to 0.1%.

### Preservatives

If a multidose formulation is required, an antimicrobial agent, called preservative, is required to be incorporated into the formulation. The preservative effectiveness must comply with the USP requirements to be qualified as multidose. The most often used preservatives and respective concentrations are phenol (0.3 to 0.5%), chlorobutanol (0.3 to 0.5%) and benzyl alcohol (1.0 to 3.0%). Additional details are provided in Chapter 5.

### Surfactants

Judicious selection of surfactants can result in the prevention of aggregation and stabilization of proteins (2). Polysorbate 80, poloxamer 188, and pluronic 68 have been used in injectable formulation. The purity of the surfactant may have an impact on the chemical stability of the preformulation. Peroxide residues in the surfactant have been implicated in oxidations of protein.

## Glass Vial Selection

Type I glass, as classified in the USP, should be used. The selection of a glass vial must also be taken into consideration when dealing with adsorptive properties of the respective protein. Adsorption of proteins will be treated later in this volume.

## Rubber Stopper Selection

In studying both the liquid and reconstituted protein drugs, the selection of a rubber stopper is also of major concern considering the potential reactivity of a protein solution with a rubber stopper, as well as the reactivity of the reconstituted lyophilized solution during storage conditions prior to use. For parenteral formulations, the biopharmaceutical industry has been using rubber stoppers with a very thin film of various inert polymers in order to achieve greater compatibility, flexibility, low levels of particulates, and machinability. In addition, adsorption, absorption, and permeation through the stopper are essentially eliminated. Extensive details may be found in Chapter 8.

## Membrane Filter Selection

Membrane filtration is the most often used technique to sterilize protein solutions. The chemical nature of the filter and the pH of the protein solution are the two most important factors affecting the protein adsorption (3). However, there are other issues that require consideration. The formulation scientist must be aware of particles or fibers released during the filtration, the potential extractables that may occur, the potential toxicity of the filter media and the product compatibility with the membrane. Of all the filters tested (unpublished data) polyvinylidene difluoride, polycarbonate, polysulfone, and regenerated cellulose were found to be the most compatible with various proteins and with minimal amounts of protein binding and deactivation.

## III. MECHANICAL AND PHYSICAL STRESSES

### A. Shaking Effect on Protein Solution at the Preformulation Level

Some of the various physical modes of vialed protein solution can undergo begin with the bulk active formulation, filling of formulated solution into vials or ampules, visual inspection, labeling, packaging, shipping, and receiving. Simulation of some of the functions described above need to be performed by doing some shaking experiments to determine their affect on aggregation induction.

Some of these preformulation experiments should also contain varying concentrations of surfactant(s) with appropriate controls. These short and inexpensive experiments can be set up on reciprocal shakers for periods of time from 1 to 6 to 24 hours, shaking from 10, 30, and 60 reciprocal strokes per minute. Reciprocal strokes disrupt and break up the flow of the liquid, while rotary strokes move the liquid circularly without breakup. These studies are intended to determine precipitation and aggregation effects. Detailed aggregation experiments and results will be described later in these chapters.

## B. Freeze–Thaw Experiments

These experiments will also be described in later chapters and will be part of Chapter 5. These experiments require a fair amount of active drug substance as well as a fair amount of work. At this point of development there may not be enough active drug substance available.

## C. Filling Systems

Of all the filling types employed to dispense liquid, such as time-pressure, piston, and rotary pump, the rolling diaphragm metering pump is the one of choice for filling biopharmaceutical solutions. The internal parts of the pump do not come in contact with one another where the liquid solution flows. This is the "TL Systems Rolling Diaphragm Liquid Metering Pump" (4). One of the most important features of this pump is that it eliminates the principal cause of particulate generation which is most often induced by parts coming together creating shedding of microscopic particles.

There are three other important parameters to control while dispensing protein solutions. (1) The speed at which liquid is filled into the vials. With protein solutions the maximum speed is between 25 to 30 vials per minute, delivering 0.5 to 2.0 mL volume per 5- or 10-mL vial per single filling head. If a large number of vials need to be filled, this filling system can accommodate variable numbers of filling heads, thus allowing it to fill a large number of vials. Filling at a faster rate will result in protein precipitation and aggregation. (2) The inner diameter of the filling cannula should not be so very small as to induce shearing and aggregation of the protein solution. (3) The tip of the cannula for the filling head should be bent at such an angle as to deliver the fluid against the inner wall of the vial and not perpendicular to the bottom of the vial. This will result in a gentle flow touching the inner wall of the vial when the cannula enters the vial and delivers the required amount of fluid. The proper bend on the tip of the cannula may also eliminate aggregation and/or shearing of the protein solution.

## D. Stability Evaluation

The development of acceptable analytical methods while isolation, characterization, and purification of a bulk active drug substance are going on is very important. It can be an aid in generating semiquantitative and quantitative measurements of the active bulk drug at various stages of the process.

Significant marketing advantages in this competitive pharmaceutical market would be to achieve a longer shelf life of the product and storage temperature at room temperature. Today the lyophilized protein drug offers refrigerated temperature storage between 2 and 8°C.

The present storage conditions set up by the USP on storage requirements are as follows:

- Cold storage.  Any temperature between 2 and 8°C
- Cool.  Any temperature between 8 and 15°C
- Room temperature.  Temperature prevailing in a working area
- Controlled room temperature.  Temperature controlled thermostatically between 15 and 30°C
- Excessive heat.  Temperature exceeding 40°C

Table 3 summarizes the initial guideline time points and temperatures that preformulation solutions should be exposed to. The results from the preformulations will allow the review team to determine directions to manipulate the excipients to obtain better stability.

**TABLE 3**   Guideline for Preformulation Stability Studies

| Temperatures | Timepoint |
|---|---|
| Frozen controls (−80 and −20°C) | Reference control sample as needed |
| Refrigerated (2–8°C) | $T$ = 0, 6, 12, 24 & 48 weeks |
| | Continue if stable |
| Intermediate (20, 30, 37°C) | $T$ = 0, 4, 8, 12, 18, 24 weeks |
| | Continue if stable |
| High temperature (40, 45, 50°C) | $T$ = 0, 1, 2, 4, 8, 12 weeks |
| | Continue if stable |

## IV. DEGRADATION MECHANISMS

To predict degradation pathways of new biopharmaceuticals is very difficult. Depending on the stress conditions, each protein may react differently than another protein. As stated previously, the objectives of preformulation are to evaluate stress conditions such as pH, temperatures, and buffers and begin evaluation of some initial breakdown products. At this particular stage of development, it is necessary to have some analytical method(s) with some reliability to detect initial degradation. It is difficult to begin evaluation of degradation products without the reliability of these preliminary assay methods.

The purpose of initial preformulation studies is to begin understanding of protein instability via chemical and physical stress conditions (5). In order to stabilize potential useful pharmaceutical products, it is important to understand how proteins degrade, how they are affected by the composition of the formulation, and the effects of stability conditions. The major pathways of protein degradation are chemical and physical. Under chemical degradation, changes and modifications occur due to bond formation or cleavage, yielding new chemical entities. One or more of the following can occur: oxidation, deamidation, hydrolysis, racemization, isomerization, beta elimination, and disulfide exchange. Physical instability can occur in the form of denaturation, aggregation, precipitation, and adsorption without covalent changes.

### A. Oxidation

Oxidation of protein is perhaps one of the most common degradation mechanisms that can take place during various stages of the processing, such as fermentation, purification, filling, packaging, and storage of the biopharmaceuticals. Under oxidative stress and in the presence of trace metals, amino acids such as methionine (Met) can be oxidized to methionine sulfoxide, cysteine (Cys) to cysteine disulfide, as well as tryptophane (Try) and histidine (His) via other modifications.

Oxidation can be controlled or minimized by (1) the addition of antioxidants, (2) having strict controls on the processing operations, (3) using nitrogen gas to flood head space of the container.

Oxidized human growth hormone (hGH) retains only 25 percent the activity of the native molecule, recombinant interferon-beta loses considerable antiviral activity due to oxidation (5). Oxidation can be detected by reversed phase HPLC (RP-HPLC), high-performance isoelectric chromatography (HP-IEC), peptide mapping, amino acids analysis, and mass spectrometry (MS) (6). In terms of total protein concentration, ultraviolet spectrophotometry is the method most often used (7).

## B. Deamidation

Deamidation is another more frequent degradation mechanism affecting pharmaceutical protein stability. Deamidation is the hydrolysis of the side chains amide on asparagine (Asp) and glutamine (Gln) to form Asp and/or Gln residues. Extensive reports have elucidated mechanisms of deamidation reactions (8).

Deamidation can be detected by isoelectric focusing, ion exchange chromatography, tryptic mapping and HPLC (9).

## C. Hydrolysis

Hydrolysis is another most likely cause of degradation of proteins. It involves a peptide (amide) bond in the protein backbone (5). The most influential factor affecting the hydrolytic rate is the solution pH.

## D. Racemization

Proteins may also degrade via other modifications (10) such as racemization. This mechanism involves the removal of the alpha proton from an amino acid in a peptide to yield a negatively charged planar carbanion. The proton can then be replaced into this optically inactive intermediate, thus producing a mixture of D and L enantiomers (2). Racemization can yield enantiomers in both acidic and alkaline conditions.

## E. Isomerization

Protein degradation is also induced by isomerization. Hydrolysis of cyclic amides of asparagine, glutamine, and aspartic acid will result in isomerization. Low pH accelerates hydrolysis of asparagine and glutamine. However, high pH accelerates hydrolysis of aspartic acid and glutamic acid (2,11,12).

## F. Disulfide Exchange

Disulfide exchange may result from a degradation other than covalent modification. These reactions may include the disulfide exchange of cysteine. This reaction is base, catalyzed and promoted by thiol antioxidants (13).

Disulfide exchange can occur in misfolded conformers due to incorrect intramolecular disulfide bonds (14).

## G. Beta-Elimination

Another degradation residue can be the beta-elimination of ser, thr, cys, lys and phe residues. These reactions are accelerated by basic pH, temperature, and the presence of metal ions (16).

## V. PHYSICAL DEGRADATIONS

### Aggregation

Protein aggregation can be of a covalent or noncovalent nature (17,18).

## A. Covalent Aggregation

This pathway involves modification of the chemical structures resulting in new chemical structures and may include reactions, such as oxidation, deamidation, proteolysis, disulfide interchanges, racemization, and others.

## B. Noncovalent

This instability may be induced by agitation, shear, precipitation, and adsorption to surfaces.

## C. Aggregation

Protein aggregation derived from either physical or chemical inactivation, is presently a major biopharmaceutical problem (17–21). Aggregation can be either covalent or noncovalent, occurring during any phase of product development from purification to formulation. An early detection of aggregation via biochemical or spectrophotometric methods, or both, can be of significant guidance to formulation scientists in selecting compatible excipients to minimize and/or prevent its formation in the experimental formulation.

Formation of aggregation can begin by the formation of initial particles from protein molecules via the Brownian movement. This is followed by collision of these molecules and aggregates of varying sizes can be formed. These aggregates can be generated by shear or collisional forces (22).

Detection and measurements of aggregations can be performed by a number of techniques. Visual observations, light scattering, polyacrylamide gel electrophoresis, UV, spectrophotometry, laser light diffraction particulate analysis, fluorescence spectra and differential scanning colorimetry (DSC), RP-HPLC, and SE-HPLC (7,14). Conformational changes can also lead to aggregations and can be measured by DSC (23).

A formulation scientist should focus on some important observations that need to be made to answer some potential problems on aggregation.

- Determination of initial approximate number of aggregates
- Determination of approximate size and distribution of aggregates
- Do the aggregates increase in size and number over time?
- Do the aggregates affect the efficacy of the proteins?
- What is the effect of aggregation on the long-term storage of the potential marketable product?

## D. Denaturation

Denaturation of proteins can be the result of several processes and reported by several investigators (27).

Factors which induce denaturation are heat or cold, extreme pHs, organic solvents, hydrophilic surfaces, shear, agitation, mixing, filtering, shaking, freeze–thaw cycles, ionic strength, and others. Thermal inactivation processes will induce conformational side reactions and destruction of amino acids (28). The loss of biological function may well be attributed to the effect of the temperature on the higher-ordered structure of the protein.

Thermal denaturation of proteins is of great interest to the formulation scientist. Thermal probes offer tools to study protein structure and stability that ultimately can be of significant use to stabilize protein drug formulations. Modifications of protein thermal effects have been reviewed (8).

The ability of the protein to refold from a denatured state, a reversible heat denaturation, is also of considerable interest for the stability of a protein formulation. These processes of renaturation are very complex (29), and each protein does have its own unique renaturation mechanisms.

Since filtrations and volume reductions occur from the fermentation to process purification, there is very likely inactivation of the protein attributable to shearing effect.

## E. Precipitation

Precipitation in formulations can occur by a variety of mechanisms such as shaking, heating, filtration, pH, and chemical interactions. Aggregation is the initial onset of precipitation. The protein molecules form aggregations of varying sizes first, and later when the aggregates reach a critical mass, precipitate out of solution and are clearly visible.

From a biopharmaceutical formulation point of view, precipitation can occur in membrane filters, filtration equipment, pumps, and tubing and loss of activity is very often recorded.

Eventually, as the aggregation mechanisms are controlled and prevented, precipitation is essentially reduced or avoided. Details on the functions of stabilizers are discussed later in this volume.

## F. Adsorption

Some of the most prevalent, ubiquitous factors of deactivation (30,31, 32) that the protein biochemists and formulation scientists face, are the surface areas interactions from the purification, formulations, and stability stages.

Essentially, at each point that the protein solution has encountered air during mixing (process), filtration (process), and air in the process steps, a significant surface area has been encountered to yield interphases.

During the actual final manufacturing of vials. ampules, syringes, catheters, pumps, and their respective storage conditions, the proteins could be adsorbed at the interphase and removed from the solution.

Several researchers (22,33,34,35,36) have investigated these biochemical mechanism problems. Since proteins have surfactant characteristics, they have a high affinity to adsorption at the air–liquid and solid–liquid interphase. Hydrophobic and hydrophilic interactions which are concentration dependent, determine the extent and the rate of adsorption. The adsorption effect on the protein is the unfolding of the protein. When this occurs at an interphase, it can lead to (1) inactivation of the protein solution, (2) insoluble protein aggregates being formed at the adsorbed site, (3) additional conformational changes occurring, and (4) chemical degradation of the protein continuing during stability periods.

## VI. SUMMARY

The initial critical parameters of preformulations have been addressed in this chapter. The formulation team, at this point of development, will review and evaluate all the results obtained from the preformulation studies. The pharmaceutical formulator will design several approaches for the next stage of formulation development taking into account all the parameters that may achieve one or more stable marketable formulations. In the formulation studies ahead, a number of stabilizing ingredients should be considered to achieve acceptable industrial stability.

## REFERENCES

1. Glynn LG. J Parenter Drug Assoc 1980; 34:139.
2. Henson FA, et al. J Colloid Interface Sci 1970; 32:162.
3. Hansen AM, Edmond Rowan SK. In vivo pathways of degradation and strategies for protein stabilization. In: Ahern JT, Manning M, eds. Stability of Protein Pharmaceuticals, Part B. Plenum Press, 1992.
4. TL Systems Corp, 5617 Corvallis Ave, North Minneapolis, MN 55429.
5. Shihong L, et al. Pharmaceutical News 1995; 2:12.
6. Mayers LC, Jenke RD. Pharm Res 1993; 10:445.
7. Sluzky V, et al. Pharm Res 1994; 11:485.
8. Chemical and physical pathways of protein degradation. In: Ahern JT, Manning M, eds. Stability of Protein Pharmaceuticals, Part A. Plenum Press, 1992.
9. Senderoff CS, et al. Pharm Abst 1993; 10:5–90.
10. Nishi K, et al. Peptide Chemistry. Japan: Protein Research Foundation, 1980:175.
11. Fischer G, et al. Biochem Biophys Acta 1984; 87:791.
12. Bachinger HP. J Biol Chem 1987; 262:17144.
13. Kenney J, et al. Lymph Res 1986; 5:23.
14. Constantino RH, et al. Pharm Res 1994; 71:21.
15. Kenney WC, et al. Lymph Res 1986; 5:23.
16. Shihong L, et al. Pharm News 1995; 2:12.
17. Klibanov AM. Adv Appl Microbiol 1983; 29:1.
18. Manning MC, Patel K, Borchardt RT. Pharm Res 1989; 6:903.
19. Weiss M. Genetic Engineering News 1994, Jan.
20. Creighton TE. Proteins: Structures and Molecular Properties. New York: WH Freeman, 1984.
21. Scopes RK. Protein Purification Principles and Practices. 2d ed. Berlin: Springer-Verlag, 1987.
22. Glatz EC. Chemical and physical pathways of protein degradation. In: Ahern JT, Manning M, eds. Stability of Protein Pharmaceuticals, Part A. Plenum Press, 1992.
23. Dingledine M, et al. Pharm Res Abstracts 1993; 10:S-82.
24. Watson E, Kenney WC. J Chromatogr 1988; 436:289.
25. Quinn R, Andrade JD. J Pharm Sci 1983; 72:1472.
26. Cantor CR, Timasheff SM. The Proteins. Vol. 5. 3d ed. New York: Academic Press, 1982:145.
27. Shirley AB. Chemical and physical pathways of protein degradation. In: Ahern JT, Manning M, eds. Stability of Protein Pharmaceuticals, Part A. Plenum Press, 1992.
28. Volkin BD, Middaugh ER. Chemical and physical pathways of protein degradation. In: Ahern JT, Manning M, eds. Stability of Protein Pharmaceuticals, Part A. Plenum Press, 1992.
29. Jaenicke R. Prog Biophys Molec Biol 1987; 49:117.
30. Mizutani T. J Pharm Sci 1980; 69:279.

31. Pitt A, J Parent Sci and Technol 1987; 41:111.
32. Trusky GA, et al. J Parent Sci and Technol 1987; 41:181.
33. Hiemenz PC. Principles of Colloid and Surface Chemistry. 2d ed. New York: Marcel Dekker, 1986.
34. Parfitt GD, Rochester CH. Adsorption from Solution at the Solid-Liquid Interphase. Academic Press, 1983.
35. Engelman et al. Ann Rev Biophys Chem 1986; 15:321.
36. Jacobs RE, White SH. Biochem 1989; 28:3421.

# 5

# Formulation Development

JOHN A. BONTEMPO

Biopharmaceutical Product Development, East Brunswick, New Jersey

## I. FORMULATION REQUIREMENTS

*Preformulation Evaluation*   From the preformulation studies, there should
be some key parameters that can be of significant aid in the designs of
experimental formulations. These key parameters are (1) Initial compatibility testing of the active drug substance with some excipients, (2) Effect
of stability factors such as temperature, light, packaging components, (3)
Initial degradation products in the preformulation, and (4) the performance of stability assays for the preformulation.

     The following are some of the major considerations to be taken into
the experimental formulation designs:

## A. Characterization, Homogeneity, and Reproducibility of the Bulk Active Drug

Characterization, homogeneity, and lot-to-lot reproducibility of the bulk
active drug substance is of paramount importance. At this stage of dosage
form development, a great deal of characterization of the bulk has been
obtained. Regulatory compliance demands that the process in place yields
reproducibility of the active drug substance, as well as whatever impurities

may be present. What is important is that they are quantitatively reproducible from lot-to-lot and that whatever impurities found have no toxicological and biological effects on the host.

## B.  pH Effect on a Formulation

The pH has a critical impact on formulations of proteins and peptides. It has a solubility and a stability impact on the formulations. With a monoclonal antibody at pH 4.2 in two different buffers, there was significant degradation as opposed to pHs 5.2 and 6.7—an optimal pH range for further development (1). At these higher pHs, further formulation development was pursued.

Some peptides can be formulated at acidic pH 2.5–4.5; however, at low pH, deamidation of asparagine and glutamine occurs. At higher pH, however, oxidation of methionine, cysteine, and tryptophan can occur, as well as other degradative mechanisms (2). The optimal pH is essential for better stability.

A change of one pH unit will change the reaction one way or another. The solution pH may be one of the most effective ways to stabilize a liquid formulation (3).

## C.  Stabilizers Used in Protein Formulations

Degradation of proteins can be a major biopharmaceutical problem during purification, characterization, preformulation, formulation development, and possibly during storage. Selective excipients are incorporated into the formulation in order to improve the physical and chemical stability of the protein drug substance.

A variety of molecules have been used as stabilizers, such as surfactants, amino acids, polyhydric alcohols, fatty acids, proteins, antioxidants, reducing agents, and metal ions. Some of the most often used excipients are stabilizers, and an explanation for their mode of action has been reported in the literature and listed in Table 1 (4–19).

## D.  Surfactants

Protein surfactant interactions have also been investigated by other researchers (20–22). Most recently, the interaction of Tween 20, Tween 40, Tween 80, Brij 52, and Brij 92 were studied with recombinant human growth hormone and recombinant human interferon gamma for surfactant:protein binding stoichiometry.

**TABLE 1**   Stabilizers Used in Protein Formulations

| Stabilizer | Action/uses |
| --- | --- |
| Proteins | |
|   Human serum | Prevents surface adsorption |
|   Albumin (HSA) | Conformational stabilizer |
| | Complexing agent |
| | Cryoprotectant |
| Amino acids | |
|   Glycine | Stabilizer |
|   Alanine | Solubilizer |
|   Arginine | Buffer |
|   Leucine | Inhibit aggregation |
|   Glutamic acid | Thermostabilizer |
|   Aspartic acid | Isomerism inhibitor |
| Surfactants | |
|   Polysorbate 20 & 80 | Retard aggregation |
|   Poloxamer 407 | Prevent denaturation |
| | Stabilize cloudiness |
| Fatty Acids | |
|   Phosphotidyl choline | Stabilizer |
|   Ethanolamine | |
|   Acethyltryptophanate | |
| Polymers | |
|   Polyethylene glycol (PEG) | Stabilizer |
|   Polyvinylpyrrolidone (PVP) 10, 24, 40 | Prevent aggregation |
| Polyhydric alcohol | |
|   Sorbitol | Prevent denaturation |
| | Aggregation |
|   Mannitol | Cryoprotectant |
|   Glycerin | May act as antioxidant |
|   Sucrose | |
|   Glucose | Strengthen conformational |
|   Propylene glycol | Prevent aggregation |
|   Ethylene glycol | |
|   Lactose | |
|   Trehalose | |
| Antioxidants | |
|   Ascorbic acid | Retard oxidation |
|   Cysteine HCl | |
|   Thioglycerol | |
|   Thioglycolic acid | |
|   Thiosorbitol | |
|   Glutathione | |

TABLE 1 Continued

| Stabilizer | Action/uses |
|---|---|
| Reducing agents | |
| Several thiols | Inhibit disulfide bond formation |
| | Prevent aggregation |
| Chelating agents | |
| EDTA salts | Inhibit oxidation by removing |
| Gluthamic acid | metal ions |
| Aspartic acid | |
| Metal ions | |
| $Ca^{++}$, $Ni^{++}$, $Mg^{++}$, $Mn^{++}$ | Stabilize protein conformation |

This stoichiometric relationship can be applied to protein formulations to determine stability. Poloxamer 407 (Pluronic F-127) was also tested with interleukin-2 and urease resulting in increased stability when the formulation was subjected to strong agitation (23). Recombinant urokinase losses were reduced by the addition of human serum albumin (HSA), Tween 80, and Pluronic F-68 (24).

Interleukin-2 and ribonuclease A, when reconstituted with a variety of surfactants, amino acids, sugars and other substances, reduced aggregation significantly (25).

The formation of particulates with a monoclonal antibody was inhibited by Tween 80 and recorded by visual and laser light diffraction particulate analysis methods (26).

Proteins will adsorb at interphases such as liquid/air or liquid/solid. When protein molecules are adsorbed they undergo physicochemical changes. Insoluble particles begin to form, eventually resulting in aggregation and precipitation and this, in turn, may lead to partial or full loss of bioactivity.

The addition of surfactants poloxamer 188 (Pluronic 68), or polysorbate to a liquid formulation can prevent or reduce denaturation of the protein at a liquid/air or liquid/solid interface of the protein in solution (27).

The most recent literature concerning the use of nonionic surfactants, indicate that during bulk storage and usage, hydroperoxides may be formed and can degrade many proteins (28).

It is for this reason that, when these surfactants are purchased, a client must ask the vendor for a certificate of analysis specifying all the tests performed, including hydroperoxides.

## E. Buffer Selection

The primary objective in selecting suitable buffers is that the buffer should have considerable buffering capacity to maintain the pH of the product at a stable value during storage condition in its marketed final container. These should be physiological buffers, USP approved. The ionic strength should also be taken into consideration since it can affect stability and isotonicity and when administered intramuscularly, the higher the ionic strength, the higher the pain in situ. In Table 2, U.S.P. physiological acceptable buffers for parenteral administration are listed.

## F. Polyols

Polyols are substances with multiple hydroxyl groups, including polyhydric alcohols and carbohydrates. These include mannitol, sorbitol, and glycerol. These have been found to stabilize proteins in solution in varying concentration from 1.0 to 10%. Although the mode of action of protein stabilization is not yet clear, it is suggested that the sugar exerts pressure to reduce the surface contact between the protein and the solvent (29,30).

## G. Antioxidants

Oxidation is one of the major factors in protein degradation. A protein solution, from purificatioin to final product for an end user, goes through various equipment made of metal, glass, or plastic. At some points during the process, the protein solution comes in contact with catalyzing metals such as copper, iron, calcium, and manganese, thus inducing the potential loss of protein activity. A probable solution to this problem will be the incorporation of a compatible antioxidant in the formulation. Some of the most often used antioxidants for parenteral preparations are ascorbic acid, sodium bisulfite, sodium metabiosulfite, monothio-glycerol, alpha tocopherol, and others. The most frequently used concentrations are in the 0.1% range and higher. The optimal concentrations are determined by the data the formulator obtains from experimental results on a case by case evaluation. Nitrogen and argon gas is also used to retard or prevent oxidative reactions and the gas is used by flooding the head space of a vial or ampule during sterile filling.

Antioxidants fall into one or more of the following categories (31):

1. *Chelating agents.*   Oxidative reactions catalyzed by metal ions. Chelating agents such as EDTA and citric acid decrease their effectiveness.

**TABLE 2** USP Compatible Buffers for Parenteral Use

| Buffering agent | KA Values (PKA) | Approximate buffering range |
|---|---|---|
| **Monobasic acids** | | |
| Acetic | $1.8 \times 10^{-5}$ (4.8) | 3.8–5.8 |
| Benzoic | $6.5 \times 10^{-5}$ (4.2) | 3.2–5.2 |
| Gluconic | $2.5 \times 10^{-4}$ (3.6) | 2.6–4.6 |
| Glyceric | $2.8 \times 10^{-4}$ (3.55) | 2.6–4.6 |
| Lactic | $8.4 \times 10^{-4}$ (3.1) | 2.1–4.1 |
| **Dibasic acids** | | |
| Aconitic | (1) $1.58 \times 10^{-3}$ (2.8) | 2–5.5 |
| | (2) $3.5 \times 10^{-5}$ (4.46) | |
| Adipic | (1) $3.9 \times 10^{-5}$ (4.41) | 3.4–6.3 |
| | (2) $5.29 \times 10^{-6}$ (5.28) | |
| Ascorbic | (1) $6.76 \times 10^{-5}$ (4.17) | 3.2–5.2 |
| | (2) $2.51 \times 10^{-12}$ (11.6) | |
| Carbonic | (1) $4.3 \times 10^{-7}$ (6.4) | 5.4–7.4 |
| | (2) $5.6 \times 10^{-11}$ (10.3) | |
| Glutamic | (1) $7.4 \times 10^{-3}$ (2.1) | 2–5.3 |
| | (2) $4.9 \times 10^{-5}$ (4.3) | |
| Malic | (1) $3.0 \times 10^{-4}$ (3.4) | 2.4–6.1 |
| | (2) $7.8 \times 10^{-6}$ (5.1) | |
| Succinic | (1) $6.9 \times 10^{-5}$ (4.2) | 3.2–6.6 |
| | (2) $2.5 \times 10^{-6}$ (5.6) | |
| Tartaric | (1) $1 \times 10^{-3}$ (3.0) | 2.0–5.3 |
| | (2) $4.55 \times 10^{-5}$ (4.3) | |
| **Polybasic acids** | | |
| Citric | (1) $8.4 \times 10^{-4}$ (3.14) | 2.1–6.2 |
| | (2) $1.7 \times 10^{-5}$ (4.8) | |
| | (3) $6.4 \times 10^{-6}$ (5.2) | |
| Phosphoric | (1) $7.5 \times 10^{-3}$ (2.1) | 2–3.1 |
| | (2) $6.3 \times 10^{-8}$ (7.2) | |
| | (3) $2.2 \times 10^{-13}$ (12.7) | 6.2–8.2 |
| **Bases** | | |
| Ammonia (ammonium chloride) | $5.6 \times 10^{-10}$ (9.25) | 8.25–10.25 |
| Diethanolamine | $1.0 \times 10^{-9}$ (9.0) | 8.0–10.0 |
| Glycine | $1.7 \times 10^{-10}$ (9.8) | 8.8–10.8 |
| Triethanolamine | $1 \times 10^{-8}$ (8.0) | 7.0–9.0 |
| Tromethamine (Tris, Tham) | $8.3 \times 10^{-9}$ (8.1) | 7.1–9.1 |

2. *Reducing agents.* These are reducing substances and inactivate oxidizing agents. Some of the reducing agents used in pharmaceuticals are sodium bisulfite, thioglycerol and ascorbic acid.
3. *Oxygen scavengers.* These compounds are more readily oxidized than the substance they are supposed to protect, thereby preferentially reducing the amount of oxidant in solution. These are ascorbic acid and sodium bisulfite.
4. *Chain terminator.* Oxidation reactions occur via free radical procedure. Chain terminators such as thiols (cysteine and thioglycerol) react with radicals in solutions to produce a new species which does not reenter the radical propagation cycle.

## H. Antimicrobials (Preservatives)

For the development of multidose formulations, it is mandatory that an antimicrobial agent is selected and incorporated into the formulation. These antimicrobial agents are called preservatives and their function is to kill or inhibit growth of bacteria and fungi that could be accidentally introduced into a vial in the process of withdrawing dosages from the vial, thus rendering the solution adulterated. The most common preservatives used in pharmaceutical and biopharmaceutical injectable products are phenol, benzyl alcohol, chlorobutanol, metacresol, and parabens. The formulator must address the following critical issues in selecting the proper preservative:

- Antimicrobial activity
- Use concentrations
- Solubility
- Optimum pH
- Stability
- Compatibility
- Inactivation

Each of these preservatives has its own characteristic reactivity with the drug substance, the excipients and the pH. Some of these preservatives have binding properties with several proteins (unpublished data). Several papers have been published (32–36) documenting binding of pharmaceutical substances by various preservatives. The most often used concentrations of preservatives are: phenol at 0.3%–0.5%, parabens-methylparaben at 0.18%, propyl-paraben at 0.02%, metacresol at 0.3–0.5%, chlorobutanol up to 0.5%, and benzyl alcohols at 1.0–3.0%.

An injectable pharmaceutical substance meets the qualification of a "multidose" if it complies with the Antimicrobial Effectiveness Test, as described in the USP No. 23. If any multidose product is designed to be marketed in Europe or the Far East, it is imperative to know the exact test procedure requirements since the preservative test requirements vary in the United States, European and Far Eastern countries. Complete preservative characteristics are found in Table 3.

## I. Tonicity

The pharmaceutical scientist, as we have read thus far, must fulfill several key requirements for a successful formulation. Another key requirement to consider is tonicity. Parenteral injectables are most desirable as isotonic solutions. In controlling isotonicity, we can control tissue damage irritation, pain, hemolysis, and crenation of the red blood cells. Hypertonic solution causes shrinkage (crenation) of the red blood cells and is reversible. Hypotonic solution will cause swelling and bursting of the red blood cells (hemolysis).

To control tonicity at all times may not be possible because of the high drug concentrations and low volumes required by some injections. When necessary, tonicity modifiers such as dextrose, sodium, and potassium chloride can be used, but it is more advisable to use sugars in place of salts. Tonicity can be calculated by several methods (37).

## II. CONTAINER-CLOSURE INTERACTIONS

### A. Glass Vials

Parenteral vial containers must be designed and packaged in such a way as to maintain package integrity. It must maintain product sterility, it must be convenient for shipping and storage, and prevent leakage.

The type of glass recommended for protein formulation is the USP Type I glass because it is the most unreactive of the glasses available. In order to achieve an excellent seal with the rubber stopper, proper dimensions of the vial and the stopper are required, thus assuring a good contact. Chapter 8 of this book covers various aspects of the elastomeric closures with focus on protein interaction.

### B. Leakage Tests

Parenteral solutions in a finished vial must prevent liquid leakage either in or out of the vial. In some cases, vacuum or gas headspace need to be controlled. There are three main tests to be performed (38). These are leakage for gas, liquid, and microorganisms (39).

**TABLE 3**   Characteristics of Preservatives

|                             | Benzyl alcohol                                      | Chlorobutanol                                                                          |
| --------------------------- | --------------------------------------------------- | -------------------------------------------------------------------------------------- |
| Antimicrobial activity      | Bacteria, weak against fungi                        | Bacteria, fungi                                                                        |
| Use concentrations          | 1.0–3.0%                                             | Up to 0.5%                                                                              |
| Solubility                  | 1:25 in water                                       | Soluble in water (1:125), more soluble in hot water<br>Soluble in ethanol              |
| Optimum pH                  | 4–7                                                 | Up to 4.0                                                                               |
| Stability                   | Slowly oxidizes to benzaldehyde                     | Decomposed by alkalies                                                                  |
| Compatibility/inactivation  | Inactivated by nonionic surfactants (Tween 80)      | Incompatible with some nonionic surfactants (10% Tween 80)<br>Decomposes at 65°C       |
| Comments                    | Bacteriostatic<br>Used for parenteral and ophthalmic products<br>Local anesthetic action | Wide range of compatibility<br>Local anesthetic action<br>Widely used |

There are also three mechanical tests that need to be performed (40), namely, needle penetration, coring, and vapor transmission.

## C.  Plastic Vials

The pharmaceutical industry introduced plastic containers because of some of the advantages plastic appeared to have, including durability, easier manufacturing, more flexibility, and perhaps more biocompatibility. How-

| Metacresol | Parabens (Hydroxybenzoates: methyl, propyl) | Phenol |
|---|---|---|
| Bacteria, fungi | Primarily fungi and gram positive bacteria Poor vs. pseudomonads | Bacteria, fungi |
| 0.3–0.5% | Methylparaben 0.18% Propylparaben 0.02% | 0.3–0.5% |
| 1:50 in water | Methylparaben (in water) 1:400 Propylparaben (in water) 1:2000 Alcohol 1:2:5 | 1:15 in water |
| 2–8 | (3–8) | Wide range (2–8) |
| Activity decrease at high pH | Essentially good | Activity decreases at high pH |
| May be inactivated by iron and certain nonionic surfactants | Serum reduces activity, also nonionic surfactants | May be inactivated by iron, albumin and oxidizing agents May be incompatible with with some nonionic surfactants |
| The meta isomer is most effective and least toxic, ortho is the weakest Mode of action is apparently related to solubility in fatty portions of organisms Combine with and denature proteins | Binds to PEG Slightly soluble Stable and nonirritating Proposed to block essential enzyme system of microorganism | Mode of action is physical damage of the cell wall and enzyme inactivation by free hydroxyl group |

ever, plastic containers were found to be prone to sorption, gas permeation, and leachables (41).

## D. Sorption of Preservatives by Plastic

Interactions of preservatives with plastics have been reviewed (42). Several preservatives were studied, including benzyl alcohol paraben, benzalkonium, and benzethionium chloride with plastic materials such as polycarbonate, polystyrene, polypropylene, polyvinyl chloride, and others. There occurred 20 to 40% loss of concentration after three months stability. This could be a significant problem that warrants study case by case.

## E. Siliconization of Elastomeric Closures

Siliconization of elastomeric closures, with a 2.0% solution of Dow Corning 360, was usually necessary to give an elastomeric closure better insertion into the neck of a glass vial. High speed filling certainly required this treatment. Without it, all kinds of problems arose during manufacturing. However, with proteins and peptides, significant problems were encountered in dealing with potential adsorptive problems between the protein-silicone-elastomeric interactions (unpublished data). Silicone traces also interfered with the development of analytical methodology, for it complexed readily with the proteins.

The latest advance in closure development is the application of a very thin flexible coating of nonreactive polymer on the elastomeric closure, such as teflon. This technology improves the insertion of the elastomeric closure into the vial, gives good seal integrity, reduces particulates associated with elastomer manufacture and washing, and eliminates silicone treatment. Pharmaceutical elastomeric closure manufacturing companies are solving these problems by researching adequate and nonreactive polymers to coat their elastomeric surfaces.

## F. Siliconization of Vials

Siliconization of glass vials has been an industry practice for some time in order to achieve complete drainage of the formulation from the walls of the container. However, with biopharmaceutical products such as proteins and peptides, siliconization has generated some difficult problems. Even though the vials, after siliconization, are baked in an oven at about 250°C for about five to six hours, during stability storage, at varying temperatures, the silicone layer begins to flake off over a period of several months. When this occurs, there could be initially visible a light haze formed by the interactions of silicone residues with the formulations. In addition, the light haze interferes with quantitative analytical development (unpublished data). The use of siliconization should judiciously be determined case by case, while perhaps new research on more inert coatings are discovered to reduce adsorptive surface properties.

## III. OTHER FORMULATION CONSIDERATIONS

### A. Shake Test

Determine the amount of physical stress the formulation in the final container can withstand by using various modes and different temperatures to simulate some shipping conditions. Some of the results will be applicable to the design of an applicable and suitable shipping container.

### B. Freeze–Thaw Cycles

Again, the physical stress of freeze–thaw cycles can have significant detrimental effect(s) on the formulation compounds; therefore, as part of the shipping validation studies, the dosage form storage is simulated from –40°C or –20°C to 2–8°C. These temperature ranges are product-to-product specific.

The cycle will begin from a frozen state at –20°C or –40°C, to a 2–8°C temperature, and subsequently to room temperature, over specific time periods. A typical freeze–thaw cycle is a 24 hour period.

After each thawing, samples are taken and assayed. The samples are frozen again, and so on. The most frequent freeze–thaw cycle is 5 days. Freezing and thawing cycles can be performed with rapid or slow cooling and with slow or fast warming. Fast warming should not exceed 25°C temperature.

The amounts of dimer formation increase with the number of freeze–thaw cycles. These dimers may or may not be reversible; this is protein and formulation dependent (43).

### C. Mechanical Stressing

Physical factors that must be controlled during the formulation development and varying stability conditions are: shaking, shearing, freeze–thaw freezing rates, liquid filtration, and filling under pressure can have significant detrimental effects, such as denaturation, adsorption, and aggregation (44–46).

### D. Stability Evaluation

The objectives of stability studies are to determine, and comply with cGMPs and regulatory requirements to establish, an expiration date and the appropriate storage conditions. The stress conditions used in preformulation stability evaluation, both physical and chemical, will be of significant guidance in formulation approaches and indicate specific excipients to be used to improve stabilization and integrity of the formulations.

## E. Setting Up Potential Formulation Candidates

To ensure that at least two or three different formulations will survive the rigorous screening, leading to a desirable marketable dosage form, the formulator should design six or more final candidate formulations. These should represent varying concentrations of the active drug substance, buffers, and selected excipients in order to achieve the most stable formulation with acceptable shelf life (47–50). Presently the majority of protein drugs on the market are stored at 2–8°C for 15–18 months.

The various excipients selected for each formulation should be acceptable by regulatory agencies. This is very important because each excipient selected by a formulating scientist must be justified for its use and at the concentration selected. *More is not better in formulations.* On the contrary, it is wise to select only those ingredients that are necessary to impart desirable stability for product superiority.

## F. Points to Consider in Setting Up Stability Studies

*Analytical Assay Methodologies.*   When the formulations reach this stage of development, it is highly necessary that at least two methods of assay have been developed that are "stability-indicating" assays. One assay alone cannot be considered sufficient, and not accepted by CBER, to monitor the potential degradation products induced by chemical or environmental routes. These assays will eventually be rigorously validated to assure measurable quantities of degradants over time. Dr. Sharma, in Chapter 6, will focus on the development of these assays that will have accuracy, precision, linearity, sensitivity, show spiked recovery, potency, strength, and stability indicator.

Table 4 summarizes the characterization and control of biopharmaceuticals.

Calculate the number of vials required for each test for each specific time point, taking into consideration the following:

- Number of batches
    At least three, if possible and available
- Active drug substance.
    At least three different lots from the final process
- Batch size
    Enough for stability requirements, plus large overage (for unexpected testing and FDA requirements)

TABLE 4  Characterization and Control of Biopharmaceuticals. Methods of Biopharmaceutical Characterization and Control, Their Uses, and References

Amino acid analysis for identity, structural analysis, and quantity
Amino acid sequencing (N- and C-terminal) for identity and structural analysis
Biochemical and colorimetric assays for activity, identity, and quantity
Biosensor assays for identity, activity, and quantity
Capillary electrophoresis for quantity, purity, heterogeneity, and stability
Carbohydrate mapping, compositional, sequence and linkage analysis for
    heterogeneity and structural analysis
Cell-based bioassays for activity
Differential scanning calorimetry for stability
HPLC for quantity, purity, heterogeneity, and stability
Immunoassays for quantity, impurity, and identity
Isoelectric focusing for identity and heterogeneity
Mass spectrometry for identity, heterogeneity, and stability
Microbiological testing for impurity
Nuclear magnetic resonance for structural analysis
Peptide mapping for identity
Residual DNA analysis for impurity
Residual moisture analysis for lyophilization efficiency
SDS-PAGE for purity, heterogeneity, and identity
Spectroscopy (UV, CD/ORD, infrared, fluorescence) for quantity and struc-
    tural analysis
Ultracentrifugation (analytical) for heterogeneity, stabiloity, and structural
    analysis
Western blots for impurity
Whole animal assays for activity
Hyphenated techniques (LC-MS, CE-MS) for identity, heterogeneity, stability,
    and quantity

Reproduced with permission of Dr. Thomas J. Pritchett. BioPharm, Vol. 9, Number 6, pg. 35, 1996.

- Specifications
  Samples for both in-house and regulatory requirements

In setting up temperature stability studies, the highest temperature points require the least amount of sample. At 40–45°C, protein products are not anticipated to be stable for more than several days. However, at 2–8°C, the temperature most likely to have the longest stability for proteins, the most samples will be required.

- Control or reference samples
  Enough samples should be stored for *retest* purpose and *retained* samples. These samples are normally stored at –40°C.
- Shipping conditions
  Final Market Container. Summer and winter conditions should be considered in the design.
- Storage position
  Upright and inverted and, if enough samples are available, place the vials in a horizontal position.
- Testing frequency
  The frequency of testing will be determined by the number of samples needed for each time point. The frequency will vary from product to product.

## IV. SUGGESTED GUIDELINES FOR MAJOR STABILITY STUDIES OF FINISHED PRODUCT AND BULK ACTIVE DRUG SUBSTANCE

- A Formulation Development Stability Program is summarized in Table 5.
- A proposed ICH storage condition is summarized in Table 6.

Bulk active drug substance stability must also be performed in order to determine stability profiles at various time points and temperature. This information has direct impact on the flexibility of how long a bulk active drug substance can be stored, and at what temperature, for manufacturing purposes.

### A. Breakdown Products

During the various time points of stability at each condition selected, the specific formulation is evaluated for characteristics such as color changes, clarity, pH, moisture transfer, extractables, tonicity, binding, adsorption, potency, stopper appearance, aggregation, particulates, and container closure integrity, and for a multidose formulation test for residual preservative(s). In addition, select the most appropriate analytical methods to monitor degradation, such as SDS-PAGE, HPSEC, IEF, CZE, RPHPLC, and others, if necessary. Understanding of degradation products is very important in both the initial preIND and postIND evaluation, in terms of toxicology and other pharmacological effects.

**TABLE 5**  Formulation Development Stability Program

- Preformulation
    - Time:   0, 1W, 2W, 1M, 2M, 3M
    - Temp:   C°: 2–8°, 25°, 37°, 45°
- Experimental formulation
    - Time:   0, 1M, 3M, 6M, 9M, 12M, 18M, 24M
    - Temp:   C°: 2–8°, 25°, 37°, 45°
- Primary formulation
    - Time:   0, 1M, 3M, 6M, 12M, 18M, 24M, 36M, 48M, 60M
    - Temp:   C°: 2–8°, 25°, 37°, 45°
- Market formulation
    - Time:   0, 1M, 3M, 6M, 12M, 24M, 36M, 48M, 60M
    - Temp:   C°: 2–8°, 25°, 37°, 45°
        - Lots from this formulation can be qualified as conformity lots
- Relative humidity in percent (RH)
    - At 25°C and 30°C, use 60% RH; at 40°C, use 75% RH

W = Week, M = Month.

## B. Specifications

The development of specifications for protein and peptide drugs is a control mechanism that is capable of assuring that the purification process is in place, yielding consistency from lot to lot to lot. Specifications apply to both bulk active protein drug and the finished dosage forms to insure the integrity and safety of the product throughout its shelf life and compliance with regulatory requirements governing the product.

In designing specifications of a specific protein, and a peptide drug product, the following are the key characteristics to consider: potency, purity, identity, microbiological, sterility, and physical tests. Depending on the physicochemical makeup of the active bulk drug substance, appropriate

**TABLE 6**  Proposed ICH[a] Storage Conditions

| Temperature | Time | |
|---|---|---|
| A.  25°C/60% RH | 0, 3, 6, 9, 12, 18, 24, 36 | Months |
| B.  30°C/60% RH | 0, 3, 6, 9, 12, 24, 36 | Months |
| C.  40°C/75% RH | 0, 1, 3, 6 | Months |

[a]International Conferences on Harmonization.
Federal Register, September 22, 1994.

**TABLE 7**  Specifications for a Protein/Peptide Drug Finished Dosage Form

Test methods

1. Physical evaluation
   Appearance
   pH
   Volume/container
   Moisture (lyophilized product)
   Total protein
   Particulates (for both liquid and lyophilized formulations)
2. Potency tests
   In vitro assays
   Radioimmunoassays
   Enzyme immunoassays
   Chromatographic methods
   Bioassays (animal model or cell-line derived)
   Protein content
3. Identity
   Peptide mapping
   $NH_2$ Terminal analysis
   Western blot
   Isoelectric focusing
   SDS-PAGE
   Coomassie stain (reduced and unreduced)
   Biological activity
4. Purity
   SDS-PAGE
   Coomassie stain
   HPLC-RP
   HPLC-SEC
   HPLC-Gel filtration
   DNA contamination
   Other specifications can be included depending on the specific require-
      ment of the protein.
5. Microbiological tests
   Sterility
   Pyrogens
   Mycoplasma
6. Safety
7. Degradation assays
   SDS-PAGE
   ELISA
   HPLC
   Electrophoresis

TABLE 8 Specifications for Purified Bulk Drug Concentrate

Test methods

1. Physical evaluation
   Appearance
   pH
2. Identity
   Bioassay
   Peptide mapping
   Amino acid analysis
3. Protein potency
   Nitrogen content
   HPLC
4. Biological potency
   Specific activity
5. Purity
   HPLC
   SDS-PAGE
   CZE
   IEF
6. DNA
7. Endotoxins
8. Sterility
   Other specifications can be included depending on
   the specific requirements of the individual protein.

specifications may be required. Specifications are product to product requirements.

In Table 7, some of the most applicable specifications are identified for a finished drug form and in Table 8, for a bulk active drug.

## C. Stability—Case Studies Graphs

As previously cited in this chapter, biopharmaceutical substances have their own specific physicochemical characteristics; consequently, it is very difficult to demonstrate degradation by a single bioanalytical method of assay. In order to support true stability characteristics, the formulator must judiciously select different methods in order to demonstrate the final marketable purity and identity of the product.

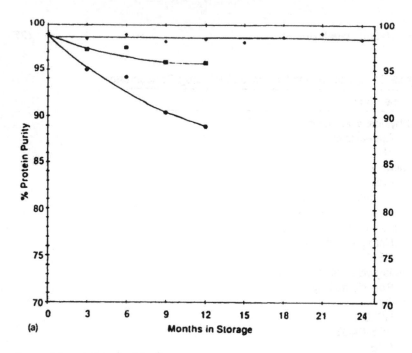

(a)

Recombinant Interleukin-2

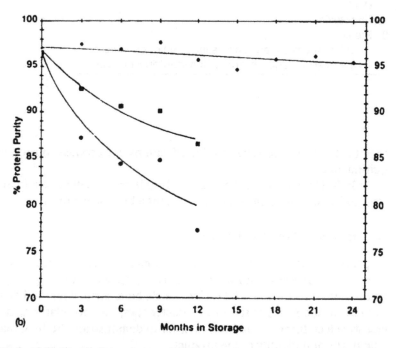

(b)

FIGURE 1   A Proleukin formulation showed decrease in purity when tested by SDS-PAGE (a) and corroborated by RP-HPLC when samples were stored at various temperatures over periods of time (b): ◆, 40°C; ■, 25°C; ●, 37°C.

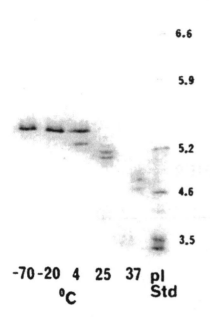

6.6

5.9

5.2

4.6

3.5

-70 -20  4  25  37  pI
       ⁰C            Std

FIGURE 2   A TNF product was undergoing deamidation during storage, as shown by the appearance of bands at pI value below 5.3; however, decomposition in SDS-PAGE method could not be detected.

Figures 1 to 8, reproduced with permission from various investigators, demonstrate the results obtained using different analytical methodologies to monitor stability.

## D. Investigational New Drug (IND) Requirements

The preparation of an IND for filing an application for a new drug to be tested in humans can be a complex, difficult, and a frustrating enterprise if the people who are responsible for preparing a specific portion of the IND have had no instruction or experience.

There are several scientific groups involved, each responsible for their specific scientific task. Table 9 represents a compilation of scientific tasks required by an IND and how long each of these tasks may take. This generic template is a summary of several INDs prepared from my experience. As it

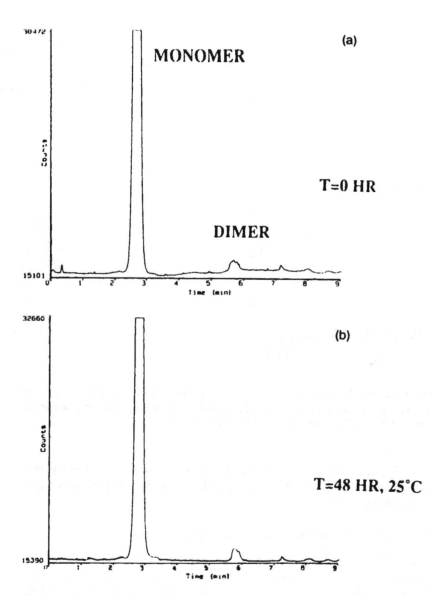

FIGURE 3 Stability samples of Proleukin (IL-2), determined by SDS-PAGE method (nonreducing) (left panel) corroborated by RP-HPLC method (right panel) immediately after reconstitution (a), and after 48 hours (b), at room temperature.

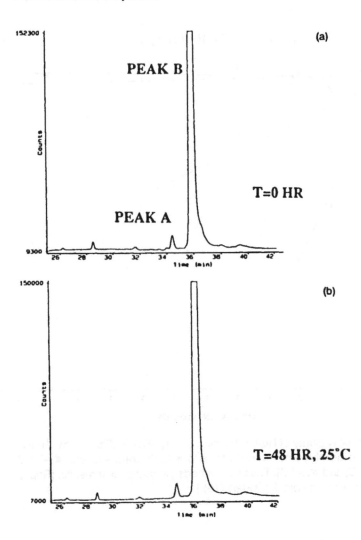

can be seen, there are several tasks that go on simultaneously to minimize the overall time requirements. When all these tasks are completed, it shows that an IND preparation may take from 9 to 12 months.

The three most critical segments are the analytical assay development, the preformulation, and formulation of the product, followed by the preclinical pharmacology.

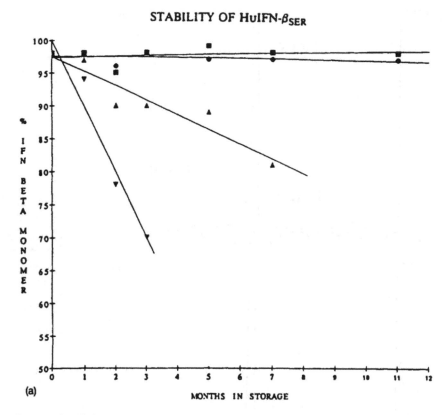

FIGURE 4    Stability purity of HuIFN-beta$_{SER}$ by (a) SDS-PAGE and by (b) RP-HPLC tested at monthly time intervals at various temperature levels (■, -70°C; ●, 4°C; ▲, 25°C; and ▼, 37°C). The two methods yielded similar results. [Reproduced with permission from J. Geigert (15).]

From the analytical development, two stability indicating assay methods should be identified. From the preformulation and formulation development, three or four formulation candidates should be identified for long range stability studies. In the present competitive market, the management of companies exert a great deal of pressure on the scientific and regulatory staff to shorten these timelines. We can be realistically aggressive and make a risky decision on a limited amount of data; however, it is when we become unrealistically aggressive that we may very well be forced to return to square one and to start all over again, very painfully.

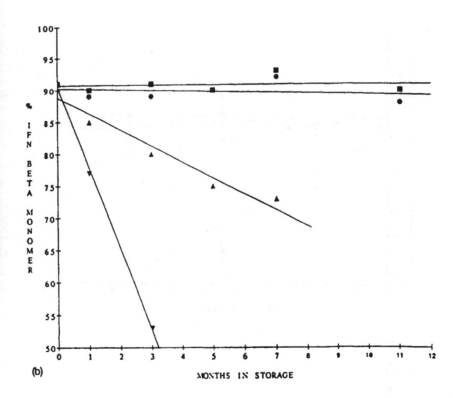

(b)                    MONTHS IN STORAGE

**TABLE 9**   IND Task/Time—Overall Requirements

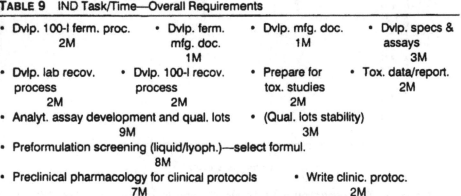

- Dvlp. 100-I ferm. proc.     • Dvlp. ferm.        • Dvlp. mfg. doc.      • Dvlp. specs &
       2M                        mfg. doc.              1M                   assays
                                   1M                                         3M
- Dvlp. lab recov.            • Dvlp. 100-I recov.   • Prepare for         • Tox. data/report.
     process                     process               tox. studies           2M
       2M                          2M                    2M
- Analyt. assay development and qual. lots           • (Qual. lots stability)
                9M                                           3M
- Preformulation screening (liquid/lyoph.)—select formul.
              8M
- Preclinical pharmacology for clinical protocols    • Write clinic. protoc.
             7M                                              2M
                    Write ind              Ind approv.
                    1.5 M                    1 M

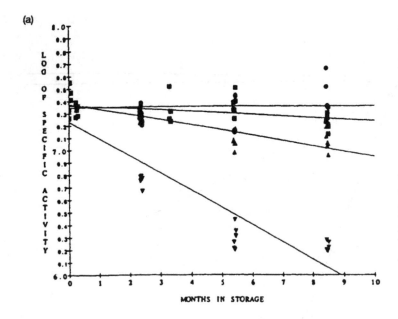

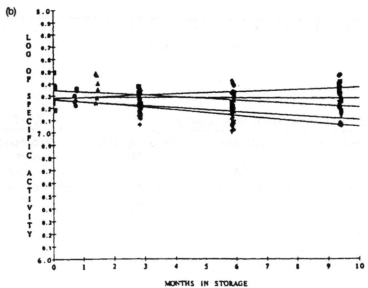

**Figure 5**  Demonstration of (a, b) stability and (c, d) purity of a liquid (a, c) versus a lyophilized preparation (b, d) of TNF. Specific activity of TNF and SDS-PAGE non-reducing gel showed similar results. ■, −70°C; ●, −20°C; ▲, 4°C; ▼, 25°C; and ♦, 37°C. [Reproduced with permission from J. Geigert (15).]

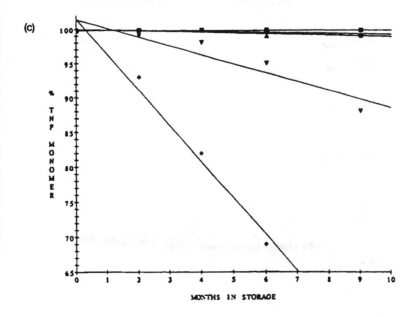

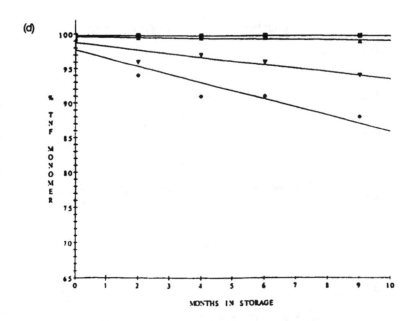

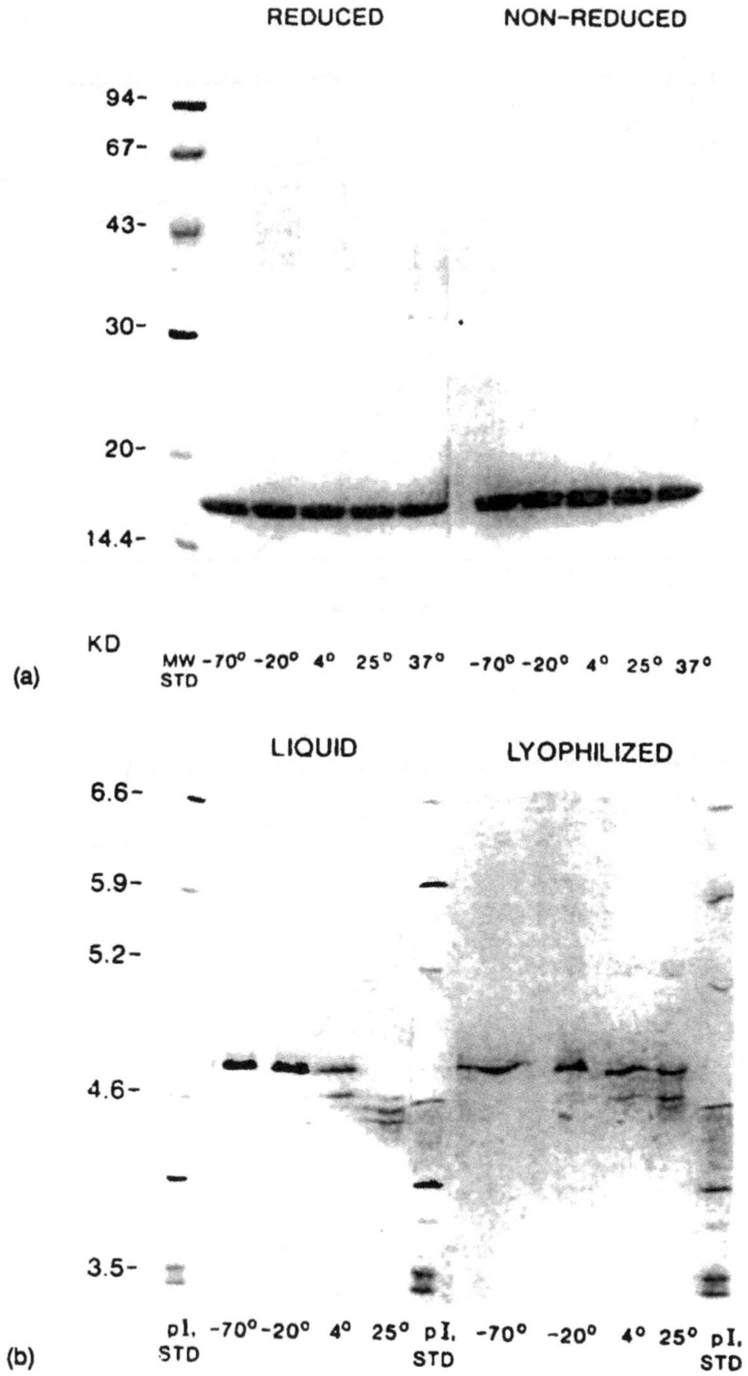

REDUCED    NON-REDUCED

94-
67-
43-
30-
20-
14.4-

KD

MW  -70° -20°  4°  25°  37°    -70° -20°  4°  25°  37°
STD

(a)

LIQUID    LYOPHILIZED

6.6-
5.9-
5.2-
4.6-
3.5-

pI.  -70° -20°  4°  25°  pI.  -70°  -20°  4°  25°  pI.
STD                    STD                    STD

(b)

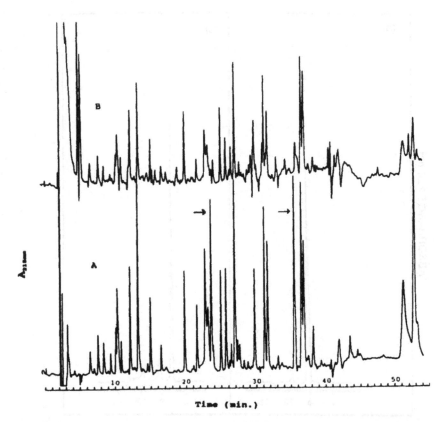

FIGURE 7   This monoclonal antibody, OKT3 peptide mapping shows chromatogram A is the standard and B is the map of the material from the
degraded sample. The results suggest that an oxidative step is involved in
the mechanism of its formulation. (Reproduced with permission from D.
Kroon.)

FIGURE 6   (a) A lyophilized sample of TNF showed no detectable deterioration
by SDS-PAGE analysis, reduced and nonreduced, or by IEF after 6 months
under various storage conditions. (b) However, after 9 months by IEF, both liquid
and lyophilized formulations showed additional bands at lower pI value, indicating the onset of deamidation, Figure 11. [Reproduced with permission from J.
Geigert (15).]

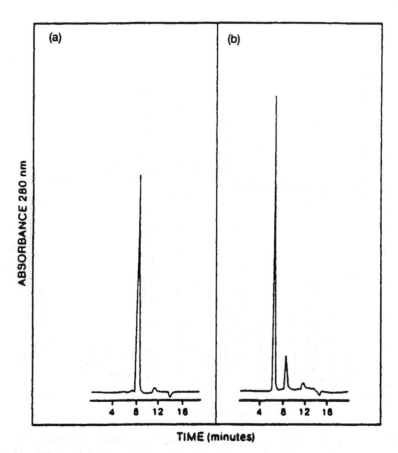

FIGURE 8  A demonstration of aggregation (a, b) and loss of activity (c) of an IL-2 preparation has occurred when exposed at 80°C for 5–8 minutes. These data demonstrate excellent correlation of aggregation and the amount remaining analyzed by HPSEC (O) and bioassay (□). From Watson, E. and Henney, W.C. (1988).

While all the IND activities are going on, additional product development issues to address are (1) research directions for scale-up, (2) preparation of preclinical supplies, and (3) preparation of initial clinical supplies.

The research directions for scale-up will require:

- Batch sheets preparation
- SOP's preparations

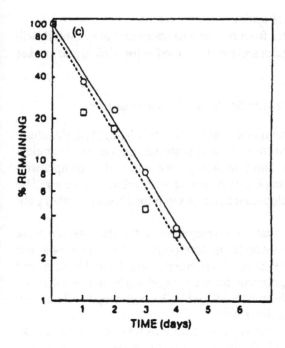

- Equipment selection and testing
- Master file preparation

The preparation of pre-clinical supplies will require dosage forms for:

- Pharmacology
- Toxicology
- Sensitization
- Irritation
- Bioburden
- Preservative efficacy
- Product stability
- GLP/GMPs compliance

Preparation of initial clinical supplies:

- Stability evaluation of the dosage forms
- Training of personnel

- Evaluate preclinical testing
- Technology transfer from research to scale-up that eventually will be transferred to pilot scale and production for clinical and market supplies.

## E. Formulation Development Scale-Up Considerations

This section of scale-up is addressed only as general highlights in this chapter. The massive documentation, such as scale-up procedures, technical documentation, validation, and final acceptance required to comply with cGMPs and Regulatory Compliance is subject matter that I am sure will be addressed by Process Development Scientists in future books or other publications, in great detail.

At this stage of formulation development, if the data demonstrate acceptable stability to warrant scale-up development, it is necessary to put the scale-up process in place. Several equipment variables will be screened and ultimately the most appropriate for the product will be selected in order to bring the product from formulated liquid bulk active drug to filled product in its final packaging configuration.

There are, however, some key considerations that do have major impact on the process. First of all, the design of the scale-up process. The process will be required to give reproducibility from lot to lot. The design of the process will reflect the choice of the final container selected, such as vials, syringes, ampules. In scaling up, another consideration of major importance will be the selection of excipients. The composition of the excipients should be the same if more than one supplier is utilized. There is, in addition, the international acceptance of the excipients and ultimately, but not last, is the cost consideration.

As covered previously, formulation and manufacturing, at this stage of development, need to focus on the physical and chemical problems associated with biopharmaceuticals. The physical adsorption of the product on surfaces such as glass, metal, and plastic and on any prefilters can have significant loss of the product, inconsistent concentration per unit container, poor yield, and ultimately rejection of the lot.

The chemical denaturation can be induced by several factors such as temperature and pressure, metal particles shedding from equipment surfaces, shear and oxidation at the air–liquid interfaces. The technology of scale-up must yield reproducible, quality attributes in the final product. When these objectives are met, we achieve a successful product development.

## F. Summary

Preformulations with biopharmaceuticals have a significant role in identifying and solving potential formulation problems. Preformulation results allow the formulation scientists to make rational designs for the experimental formulations to be tested.

Key phases of successful product development place emphasis on close, collaborative, and productive interactions of the interdisciplinary sciences within a pharmaceutical group.

Physicochemical properties of a protein–peptide drug must be identified in order to approach preformulations and formulations studies with rational designs. Selective protein stabilizers play a major role in imparting stability of the product under specific experimental conditions.

Analytical methods able to determine the potential stability or degradation products of a formulation must be developed. These methods must be validated and qualified as *stability indicators*.

The ultimate goals of the formulation and product development scientists, from fermentation to production, are to deliver to the health field a protein drug which is safe, effective, pure, stable, elegant, suitable for production, cost effective, and marketable.

## REFERENCES

1. Begum S, et al. Pharm Abst 1993; 40:S076.
2. Wang CY, Hanson AM. J Parent Sci Techn 1988; 42:53.
3. Nguyen AT. Colorado Protein Stability Conference, 1994.
4. Stewart WE, et al. Biochim Biophys Acta 1974; 359.
5. Coval ML. Injectable Gamma Globulin. US Pat 4,165,370; 1979.
6. Henson AF, et al. J Colloid Interface Sci 1970:32.
7. Coval ML. Method of Producing Intravenous Injectable Gamma Globulin. US Pat 4:124,576; 1978.
8. Quinn R, Andrade JD. J Pharm Sci 1983; 72.
9. Fukushima T, et al. Eur Pat 37078; 1981.
10. Mikaelson ME, et al. Blood 1983; 62.
11. Siegel H, et al. Chem Rev 1982; 82.
12. Busby TE, Ingham KC. Biochim Biophys Acta 1984; 799.
13. Anderson L, Hahn-Hagerdol B. Biochim Biophys Acta 1987; 912.
14. Arakawa T, Timasheff SW. Biochemistry 1982; 21.
15. Geigert J. J Parent Sci Tech 1989; 43.
16. Privalov PL, Gill SJ. Advances Protein Chemistry. New York: Academic Press, 1988:191.
17. Yang Y-CJ, Hanson MA. J Parent Sci Tech 1988; 42.

18. Akers MJ. J Parent Sci Tech 1982; 36.
19. Fujimoto S, et al. Chem Pharm Bull 1983; 31.
20. Schwuger JM, et al. In: Gloxhuber E, ed. Surfactant Sci Ser. New York: Marcel Dekker, 1980:10.
21. Greener J, et al. Macromolecules 1987; 20:2490.
22. Steinhart J, et al. Biochemistry 1977; 16:718.
23. Wang P, Johnston TJ. J Parent Sci Tech 1993; 47:123.
24. Patel PJ, et al. Pharm Abst 1993; 10:S-75.
25. Zhang Z, et al. Pharm Res 1996; 13:643.
26. Dingledine M, et al. Pharm Sci Abst 1993; 10:S-82.
27. Coval ML. US Pat 4,165,370; 1979.
28. Knepp MV, et al. J Parent Sci and Tech 1996; 50:163.
29. Lee CJ, Timasheff NS. J Biol Chem 1981; 28:7193.
30. Timasheff NS, Arakawa T. In: Creighton TE, ed. Protein Structure: A Practical Approach. Oxford, UK: Irl Press, 1989:331.
31. Connors AK, et al. Chemical Stability of Pharmaceuticals. A Handbook for Pharmacists. 2d ed. New York: J Wiley and Sons, 1986:82.
32. Lachman L, et al. J Pharm Sci 1962; 51:224.
33. Kostenbauder HB. In: Block SS, ed. Disinfection, Sterilization and Preservation. 2d ed. Philadelphia: Lea and Febinger, 1977:912.
34. Lochman L, et al. J Pharm Sci 1964; 53:211.
35. Yousef RT, et al. Can J Pharm Sci 1973; 8:54.
36. Coats D. Mfg Chem Aerosol News 1973; 44:41.
37. Martin AM, et al. Physical Pharmacy. 2d ed. Philadelphia: Lea and Febinger, 1969.
38. Morton KD, et al. J Parent Sci and Tech 1988; 42:58.
39. Morton KD, Lordi GN. J Parent Sci and Tech 1988; 42:58.
40. Smith JE, Nash JR. Elastomeric Closures for Parenteral Pharmaceutical Dosage Forms. Vol. 1. 2d ed. New York: Marcel Dekker, 1992.
41. Guazzo MD, Ambrosio JT. Container Closure System for Sterile Products: 28th Arden House Conference, 1993.
42. Kakemi K, et al. Chem Pharm Bull 1971; 19:2523.
43. Paborji M, et al. Pharm Sci Abstracts 1993; 10:S-76.
44. Mizatani T. J Pharm Sci 1980; 69.
45. Pitt A. J Parent Sci Tech 1987; 41.
46. Trusky GA, et al. J Parent Sci Tech 1987; 41.
47. Hageman MJ. Drug Dev Indus Pharm 1988; 4:2047.
48. Wang J-C, Hanson AM. J Parent Sci Tech 1988; 42:53.
49. Manning CM, et al. Pharm Res 1989; 6:903.
50. Geigert J. J Parent Sci Tech 1989; 43:220.

# 6

# The Analytical Techniques

BASANT G. SHARMA

The R. W. Johnson Pharmaceutical Research Institute, Raritan, New Jersey

## I. INTRODUCTION

The aim of this chapter is to set out the basic concepts of most commonly used analytical techniques in the pharmaceutical biotechnology industries. The attempt is made in such a way as to assist those who are embarking on the subject for the first time, or who already have some practical experience but are seeking to broaden their knowledge of the subject as a whole. It does not set out to provide an exhaustive review or to provide detailed recipes for individual assays. Instead, an attempt will be made to summarize the principle of all these assays.

The term "biotechnology" is made up of words from the ancient Greek which mean, literally, "industrial use of life forms." The pharmaceutical biotechnology industry today has the excitement and promise for the new generation of drugs, for the first time to cure or significantly alleviate untreatable major diseases. Cheaper and more effective vaccines are being developed, hormones and other factors usually produced in the body in very small quantities are now available in commercial quantities, and immune cells can be genetically engineered to increase their effectiveness in virally infected tissues. Insulin, the earliest pharmaceutical protein made available in large amounts, being derived from bovine and porcine pancreases since the 1920s, can now be made by fermentation of yeasts. The prospects of being able to transplant organs more successfully and prevent rejection of transplant organs are all very exciting developments. The subject will undoubtedly develop further as innovations designed to reduce the toll of death and suffering due to disease are introduced into pharmaceutical and medical practice. I want to emphasize that drugs derived from the application of pharmaceutical biotechnology are chemical entities that have some familiar and some unfamiliar properties when handled by the formulation and/or analytical chemist.

## A. Basic Concepts

Proteins and peptides are the principal constituents of cells; they are ubiquitous throughout nature and essential to life itself. The dominant role played by proteins in living systems can be appreciated with the selection of a few examples that illustrate the variety of physiological properties associated with these molecules. Proteins participate in such diverse and necessary processes as respiration, muscular contraction, active transport of cellular constituents, electrical transmission, and in the expression and perpetuation of genetic characteristics. All chemical reactions in living systems are catalyzed by enzymes, protein molecules that are catalysts either by themselves or when complexed with small molecules. As science develops and basic understanding increases around physiological roles of these proteins, the pharmaceutical biotechnology will develop further as innovations designed to control or regulate these functions.

Before discussing analytical techniques, it is appropriate to review fundamental aspects of proteins and peptides—their composition, physical, and chemical properties. If additional information is needed, it can be found in any standard biochemistry textbook.

Proteins and peptides are made of amino acids. Only 20 amino acids are commonly found in natural proteins, and these are combined in countless ways to form an enormous variety of different protein molecules. These amino acids can be grouped as aliphatic, aliphatic alcohol, sulfur, imino acid, acidic, amide, basic and aromatic. The three-letter abbreviation is commonly used when protein sequences are reported and the one letter abbreviations are useful when related protein sequences are compared (Table 1). Amino acids are covalently linked to one another in a protein by peptide bonds. The $\alpha$-carboxyl of one amino acid and the $\alpha$-amino of a second amino acid participate in the formation of an amide bond, linking

**TABLE 1** Amino Acid Abbreviations

| Amino acids | | Three-letter abbreviation | One-letter abbreviation |
|---|---|---|---|
| Aliphatics | Glycine | Gly | G |
| | Alanine | Ala | A |
| | Valine | Val | V |
| | Leucine | Leu | L |
| | Isoleucine | Ile | I |
| Hydroxy | Serine | Ser | S |
| | Threonine | Thr | T |
| Acidics | Aspartic acid | Asp | D |
| | Glutamic acid | Glu | E |
| Amides | Asparagine | Asn | N |
| | Glutamine | Gln | Q |
| Basics | Histidine | His | H |
| | Lysine | Lys | K |
| | Arginine | Arg | R |
| Sulfur | Cysteine | CysH | . |
| | Cystine | Cys | C |
| | Methionine | Met | M |
| Aromatic | Phenylalanine | Phe | F |
| | Tyrosine | Tyr | Y |
| | Tryptophan | Trp | W |
| Imino | Proline | Pro | P |

the two amino acids. The repeating structure of amino acids linked to one another by peptide bonds is referred to as the backbone of the polypeptide chain. The type of side chain will vary from one residue to another in the polypeptide, so that the nature of the side chain gives the protein molecule its distinctive properties and functional behavior.

The amino acid sequence of a protein, which is also referred to as the primary structure, is a complete description of the molecule in organic chemical terms, equivalent to the structural information available for small organic molecules. The primary structure of a protein provides us with a description of the organic-chemical nature of the molecule, but it fails to deal with those features of the protein that describe the conformation of the polypeptide chain.

The secondary structure of a protein is concerned with a description of the three-dimensional arrangement of the polypeptide chain—the conformation of the polypeptide backbone.

The tertiary structure of a protein refers to the orientation of the side chains in the folded molecule. The compact nature of the molecule places the side chains of amino acid residues in close physical contact with one another, close enough to promote side chain interactions. A few representative examples of side chain interactions are salt linkages and hydrogen bonding occurring between polar groups; hydrophobic and Van der Walls interactions occur between nonpolar groups. The disulfide bond is a covalent bond linking together two half-cystine residues.

Although noncovalent interactions of side chains are weak interactions, the very large number of participating groups make these forces important stabilizing factors, analogous to the hydrogen bonds of the secondary structure. It should also be obvious that the tertiary and secondary structures are intimately interrelated.

## II. ANALYTICAL TECHNIQUES

It is not practical, nor would it be particularly helpful, to attempt to give here an exhaustive review of all analytical techniques. There are several reference and text books available on this which can be referred to for additional information. The aim of this section is consistent with what was mentioned earlier—review of analytical techniques commonly used in the pharmaceutical biotechnology industries and accepted by regulatory agencies worldwide.

For simplicity, all analytical techniques commonly used in the pharmaceutical biotechnology industries can be divided into two broad categories:

separation methods and bioactivity methods. The section below will describe many separation and bioactivity methods, and their application to the field of pharmaceutical biotechnology.

## A. Separation Methods

A major aspect during characterization, stability, and purity of biotechnology products is separation of biologic components into pure fractions. For example, analysis of compounds by high-performance liquid chromatography (HPLC) involves separation of the compounds, which are identified by their characteristic elution positions and profiles and quantitated by measurement of peak areas. This not only holds true for macromolecules, but also for the analysis of small molecules. Many of the analytical separation procedures are equally effective for small molecules and large molecules. Many of them are unsuitable for macromolecules. Separation methods to be considered mainly fall into two categories: chromatography and electrophoresis. At the present time, the development of new techniques and new equipment for chromatography and electrophoresis is proceeding at a rapid rate.

## 1. Chromatographic Techniques

Chromatographic methods can be classified according to the mechanism of the separation, such as size exclusion, ion exchange, and affinity chromatography.

*a. Size Exclusion Chromatography.* Size exclusion chromatography (SEC) or gel permeation chromatography (GPC) separates molecules according to their molecular size. This method depends on the ability of a molecule to penetrate into porous solvated particles of the "stationary phase." The smaller the molecular size, the more frequently it will enter through the pores of the column material. The pores are of a defined diameter and will exclude molecules that are larger than the pores. The order of elution, then, is in descending order of molecular size. The "mobile phase" is usually aqueous. It is a relatively low-resolution method, but useful to detect low molecular weight degradative impurities or high-molecular-weight protein aggregates from protein samples.

A SEC-HPLC chromatogram is presented in Figure 1. This chromatography consisted of a TSK gel G3000SW column (60 × 0.75 cm, 10 $\mu$m, Toyosoda) and an isocratic mobile phase of 0.1 M sodium chloride in 0.02 M sodium citrate at pH 7.0, with a flow rate of 1.0 mL/min. Protein peak monitored using 280-nm wavelength.

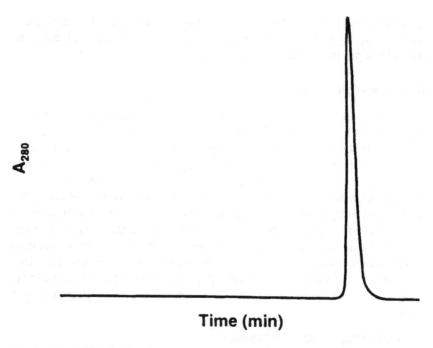

**FIGURE 1**   SEC-HPLC—chromatograph of cdrOKT4a (humanized monoclonal antibody) using TSK gel G 3000W column with an isocratic mobile phase.

Recombinant human erythropoietin is a monomeric glycoprotein hormone when stored under refrigeration. Higher temperatures and various formulation conditions have shown the monomer to partially dimerize followed by aggregation to higher molecular weight species. The method of choice is SEC-HPLC to separate and quantitate monomer, dimers and high molecular weight aggregates of erythropoietin, as recently published by DePaolis et al. (1995).

As a general rule, physical forces and chemical agents that are capable of breaking the noncovalent interactions that stabilize the protein molecule and/or stabilize aggregates must be avoided or kept to a minimum; for example, elevated temperature, high or low pH, organic solvents, and detergents that may denature a protein molecule or dissociate protein aggregates. Unfortunately, susceptibility to different conditions varies among proteins, and preliminary experiments must be performed in order to test new procedures.

*b. Ion Exchange Chromatography.* Ion exchange chromatography (IEC) separates components based on differences in their charge and, to a lesser extent, on adsorption. Ion exchange resins are high-molecular-weight polymers with functional groups that are acidic (cation exchangers) or basic (anion exchangers), while the polymeric matrix itself is stable to organic solvents and to solutions of both high and low pH. The "mobile phase" is an appropriate aqueous buffer that is varied in pH or ionic strength in order to cause movement of the sample components through the column.

Proteins are polyvalent molecules and at a given pH, different proteins have different numbers and types of charged groups that can interact with the ion exchange resin. It is this tremendous variety of possible inter-

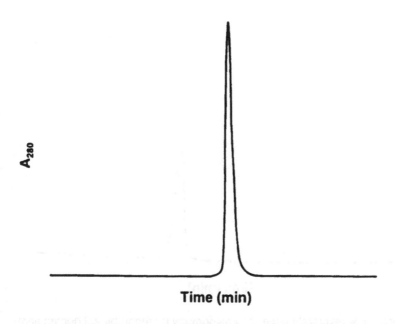

**Time (min)**

FIGURE 2  IEC-HPLC—chromatography of cdrOKT4a using cation-exchange resin. Sample is eluted by increasing ionic strength.

actions between protein and the ion exchanger that leads to the high resolving power of the technique.

A typical IEC chromatograph is presented in Figure 2. This chromatograph again shows homogeneity of cdrOKT4a. The TSK gel CM-5 PW (75 × 7.5 mm, 10 $\mu$m, Tosohaas) column as used with buffer A: 50 mm sodium acetate, pH 5.1, and buffer B: 50 mm sodium acetate with 2 M sodium chloride, pH 5.1, in a gradient mobile phase. Protein peak monitored using 280 nm wavelength.

Separations based on differences in charge can also be accomplished by electrophoresis on a supporting medium of polyacrylamide gels (see later).

*c. Reversed-Phase Chromatography.* Reversed phase chromatography (RPC) is a two-phase solvent system. One of the phases is a relatively non-

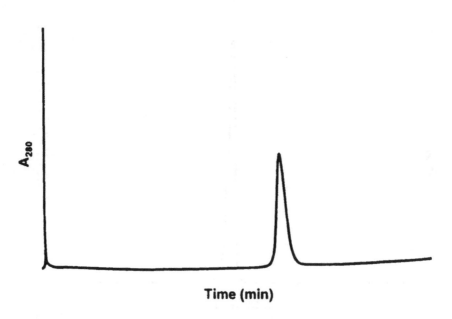

**Time (min)**

FIGURE 3 The reversed-phase chromatography for recombinant human erythropoietin (rHuEPO) consisting of C4 reversed-phase column with a gradient mobile phase.

polar organic phase, covalently bonded to a solid support and becoming the stationary phase. A relatively polar solvent, usually containing water, is used as the "mobile phase." The opportunity to use aqueous solvents and the potential for high resolution has led to many applications of this method for peptides and proteins.

A reversed-phase chromatography on erythropoietin is represented in Figure 3. This chromatography consisted of a reversed-phase C4 column (5 × 0.46 cm, 5 $\mu$m, 300 A, VydaC 214TP5405) with a gradient mobile phase of 0.06% TFA in water (A) and in acetonitrile (B). The gradient had step 1: 35 B for 5 min, step 2: 35 to 38% B in 10 min, step 3: 38 to 50% B in 20 min, and step 4: 50 to 35% B in 5 min. The flow rate was 1.5 mL/min with detection at 230 nm.

Reversed-phase chromatography is widely used for purity analyses of protein. An example with recombinant human alpha interferon (Trotta et al., 1987) clearly points out the strength of reversed-phase chromatography. This method was also used for characterization of genetically engineered Alpha-2 interferon by Nagabhushan et al. (1983) and for several other biological molecules.

*d. Affinity Chromatography.* Affinity chromatography is based on the specific binding interaction that occurs between proteins and various biomolecules such as antigens, enzyme substrates and inhibitors, coenzymes, agonist and antagonist ligands, for receptors. The column matrix may be composed of a support coupled to the protein. The major advantages of affinity chromatography are speed and specificity. The versatility of the method is increased by the use of other relatively specific ligands. Further applications are limited only by the imagination of the chemist and the properties of the individual protein.

An example of affinity column chromatography is presented in Figure 4, platelet derived growth factor (PDGF) polyclonal antibodies coupled to protein-G matrix. Elution of PDGF followed by lowering pH and monitoring protein peak at 220 nm.

All of the applications mentioned above utilize high-performance liquid chromatography (HPLC), originally known as high-pressure liquid chromatography. HPLC is very versatile and many uses have been developed that are applicable to peptides, proteins, and related aspects of developmental biotechnology. The applications which utilize reversed-phase chromatography, ion-exchange chromatography, and size exclusion chromatography are certainly well known. One of the most useful biotechnologic applications of HPLC is the separation and analysis of protein digests using proteolytic enzymes. Peptides differing in length or content by only a

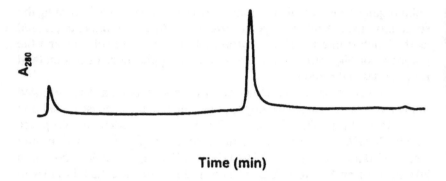

**Time (min)**

FIGURE 4   Affinity chromatography for recombinant human platelet derived growth factor (rHuPDGF-B) using polyclonal anti-PDGF antibodies coupled to support matrix of protein-G. rHuPDGF is eluted by lowering the pH.

single residue can usually be resolved readily by these techniques, using either reversed-phase or ion-exchange approaches. Individual resolved peptides of interest can be isolated easily for sequencing and further analysis.

The peptide map of the trypsin-digested light chain (L) and the heavy chain (H) of cdrOKT4a generated by reversed-phase HPLC is shown in Figure 5. Fractions containing pure H and L chain were digested with trypsin. The digests were analyzed by reversed-phase C18 column (4.6 × 250 nm, 5 $\mu$m, 300 A, Rainin Dynamax) with a gradient of increasing acetonitrile in 0.1% TFA at 1.0 mL/min. Nearly all of the peaks in the chromatogram were identified by sequencing. This helped to establish that the product had the correct amino acid sequence, as expected by grafting the complimentarily determining regions (CDRs) of murine OKT4a into a human antibody framework.

The peptide map is a very sensitive and informative method. This is widely used for investigation and understanding the mechanism of degradation of proteins. One very interesting investigation was carried out on erythropoietin by DePaolis et al. (1994) and explains the tryptic map variation of erythropoietin resulting from carboxypeptidase B-like activity. Identification of sites of degradation in a therapeutic monoclonal antibody by peptide mapping was published by Kroon et al. (1992). In summary,

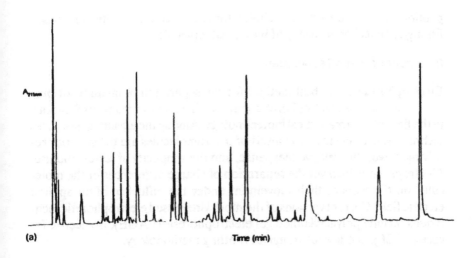

(a)                                         Time (min)

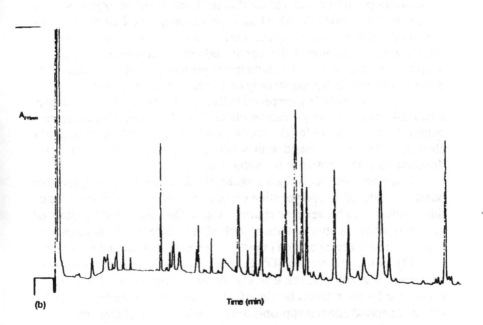

(b)                                         Time (min)

FIGURE 5   Peptide maps of cdrOKT4a light (L, 5a) and heavy (H, 5b) chains.
Separated L and H chains from cdrOKT4a were digested with trypsin and the
digests analyzed by HPLC.

peptide map is the method of choice for identification, purity, and establishing chemical equivalency of biological molecules.

## 2. Electrophoretic Techniques

Electrophoresis is the best method for the separation of mixtures of proteins. The various electrophoretic methods have found many applications in the field of pharmaceutical biotechnology. Among the advantages of these techniques over HPLC for analysis of macromolecules are the superior resolving power, the relative inexpense, and the simplicity of set-up and use. Electrophoresis involves the separation of charge differences of the molecules on the basis of their movement under the influence of an applied electric field. Current versions of the most widely used electrophoretic techniques include polyacrylamide gel electrophoresis (PAGE), isoelectric focusing (IEF), and a whole range of blotting methodology.

*a. Polyacrylamide Gel Electrophoresis.* This is probably the most versatile electrophoretic technique used in the presence of detergents such as sodium dodecyl sulfate (SDS), which is particularly useful for the determination of molecular weights by reference to a set of standards. Apparently the presence of the anionic detergent, sodium dodecyl sulfate, masks the charge differences that exist with different proteins, and the mobility of the protein is therefore dependent only on the size of the molecule.

The SDS-PAGE is widespread in its acceptance as a simple, inexpensive, rapid, and extremely sensitive method for molecular weight determination. It is also useful for the study of oligomeric proteins because the electrophoresis is performed with a detergent present, which causes the dissociation of the protein to the subunit level.

In the presence of a reducing agent, like β-mercaptoethanol, disulfide bonds are reduced and polypeptide chains separate if intermolecular disulfide bonds existed between the chains. Indeed, the behavior of a protein on gel electrophoresis after treatment with SDS or SDS and β-mercaptoethanol is a most useful approach in judging the extent of subunit structure.

Figure 6a, b represents SDS-PAGE of cdrOKT4a under nonreduced and reduced conditions, respectively. As antibodies H and L chains are connected by intermolecular disulfide bonds, after reduction H and L chains separated and run appropriately on SDS-PAGE (Figure 6b).

*b. Isoelectric Focusing.* This is the most widely accepted technique useful for the determination of molecular heterogeneity in the pharmaceutical biotechnology industries. A stable pH gradient is created between the

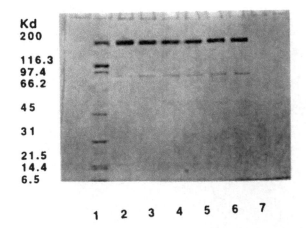

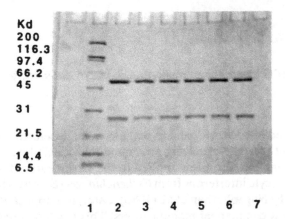

FIGURE 6  SDS-PAGE of cdrOKT4a monoclonal antibody under (a) nonreduced and (b) reduced conditions. Lane 1 is molecular weight markers. Lanes 2–7 are cdrOKT4a monoclonal antibody samples under (a) nonreduced conditions and (b) reduced conditions showing H & L chains.

anode and the cathode during isoelectric focusing. Under the influence of an applied electric field, the charge molecules in the sample move through the medium until they eventually reach a position in the pH gradient where their net charge is zero and they will migrate no further. For polypeptides and proteins, this pH is their isoelectric point (pI). Molecules with the same

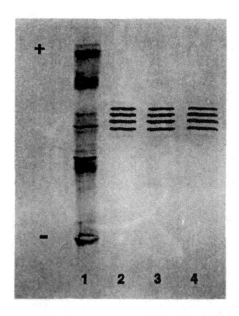

FIGURE 7   IEF of rHuEPO. Lane 1 is pI markers. Lanes 2–4 are samples of rHuEPO.

pI become concentrated in a narrow band at that pH value. Isoelectric focusing is a technique of very high resolving power and reproducibility.

Isoelectric focusing has also been applied to the purification of recombinant human leukocyte interferons from *Escherichia coli* (*E. coli*). This was a very early work by Nagabhushan et al. (1986). Later, this application was followed by several commercial manufacturers. This is cited here to help the reader understand the power of this technique.

Figures 7 and 8 represent IEF of rHuEPO and cdrOKT4a. There is no further need to question its very high resolving power and reproducibility.

*c. Western Blots.*   Blots are an extension of SDS-PAGE and are widely used in the pharmaceutical field. The proteins are transferred from the SDS-PAGE to a nitrocellulose membrane electrophoretically. After the transfer, all additional binding sites on the matrix must be blocked with excess blocking agents (nonfat dry milk or bovine serum albumin, etc.); then a specific antibody is bound and finally, a second antibody directed against the first antibody. This second antibody can be conjugated to an enzyme

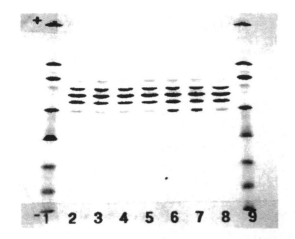

FIGURE 8    IEF of cdrOKT4a monoclonal antibody. Lanes 1 and 9 are pI markers. Lanes 2–8 are samples of cdrOKT4a monoclonal antibody.

such as horseradish peroxidase or alkaline phosphatase for immunological detection of protein on nitrocellulose membranes.

Figure 9 represents Western blot of rHuEPO. Monoclonal antibodies are used for the detection of rHuEPO as the first antibody. The second antibody, alkaline phosphate conjugated used for detection of protein on nitrocellulose membranes.

*d. Combined Electrophoresis and Isoelectric Focusing (Two-Dimensional Electrophoresis).* SDS-polyacrylamide gel electrophoresis and isoelectric focusing are the two most powerful techniques yet devised for the resolution of the components of mixtures of proteins. Isoelectric focusing separates protein molecules on the basis of their isoelectric points, whereas SDS-polyacrylamide gel electrophoresis separates them on the basis of their molecular size. Consequently, these two techniques are ideally suited for use in combination for the two-dimensional separation of proteins.

The native proteins are first separated by isoelectric focusing in polyacrylamide gel and are then denatured in situ by incubating this gel in buffer containing SDS. The denatured proteins are transferred to an SDS-polyacrylamide slab gel, and electrophoresis is carried out. Usually, gels composed of a gradient of polyacrylamide concentration are used for the electrophoresis, to give a better spread of separated components.

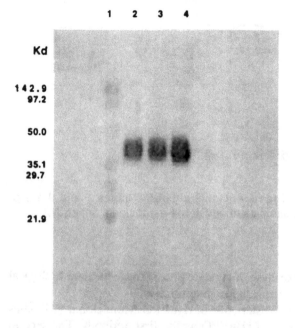

FIGURE 9   Western blot of rHuEPO. Lane 1 is prestained molecular weight markers. Lanes 2–4 are rHuEPO samples.

Application of two-dimensional electrophoresis is limited to investigational or developmental work. There are several problems associated with the reproducibility of band patterns (between samples, between experiments, and interlaboratory variation) and identification of individual components become much more demanding because by using high resolution steps for both dimensions, it is possible to resolve several protein zones on a single gel. It is also not suitable for the quantitative analysis of proteins in a sample. There is often significant loss of proteins, particularly those of lower relative molecular mass, when the IEF gel is treated with SDS, because of diffusion of the proteins out of the gel.

A number of recent reports exist on the characterization of proteins by capillary SDS gel electrophoresis (SDS-CGE) using various molecular sieving matrices. Wu et al. (1992), Ganzler et al. (1992), Werner et al. (1993), and an extensive comparison of the separation of proteins by SDS-PAGE and SDS-CGE has been published (Guttman et al., 1994). Generally, these separations are performed with non-cross-linked polymer networks that are replaced in the capillary between each electrophoresis run. SDS-CGE has several advantages over SDS-PAGE, including direct on-column quantitation by UV (or fluorescence) detection as opposed to densitometer scanning of dye-stained slab gels, speed for single-sample analysis, sensitivity, capability of automation, and ease of computerized data storage and processing by standard chromatography software instead of the commonly used photography of slab gels. Recently, Kroon et al. (1995) investigated purity determination of immunoglobulins with emphasis on the quantitation of the amount of half-antibody in genetically engineered monoclonal antibodies.

All of the electrophoresis techniques mentioned above have found many biochemical applications at various stages of product development.

## B. Bioactivity Methods

A second major aspect of pharmaceutical biotechnology product development, besides separation of macromolecule components, is the performance of appropriate bioactivity (potency) assays. The selection and design of any assay often embodies the most important elements of scientific methods, including the formulation and testing of hypotheses.

Ideally, any potency assay should be (1) relevant to clinical use; (2) easy or convenient to perform, enabling a maximum number of determinations with a minimum of manipulation; (3) reproducible, exhibiting an interassay variation of less than ten percent; (4) accurate, reflecting the true quantities of the substance and its actual fluctuations in experimental or physiologic processes; (5) precise, with an intra-assay variation of less than five percent; (6) specific, measuring only what it is supposed to measure, relative to an appropriate standard; and (7) inexpensive.

In reality, it is very difficult to develop an ideal potency assay in the early stages of product development. However, once research using the animal model has produced information indicating the mechanism of action, then in vitro assays can be developed to use for product development.

### 1. In Vivo Whole Animal Bioassay

Most of the time a developmental project is initiated using whole animal bioassay. The validity of such bioassays, which are most important during the early stages of a biological investigation leading to the identification and therapeutic formulation of a molecule, heavily depends on a clear statement of the problems to be solved. In simple words, animal bioassay mimics biological effect in humans. Animal bioassays have a proven history of use. In almost all cases, whole animal bioassays have low precision, require a large number of animals, have long analysis times (usually weeks), and are expensive (hundreds of dollars per sample).

Bioassay is the only way to understand the effect of various factors on potency. Udupa et al. (1996) examined the possible role of tumor necrosis factor-alpha in erythropoietic suppression by endotoxin and granulocyte/ macrophage colony-stimulating factor. They have examined this suppression in exhypoxic polycythemic mice, in which a wave of erythroid cell formation was initiated by single injection of erythropoietin. The primary effect of the administration of both endotoxin and granulocyte/macrophage colony-stimulating factor was to induce tumor necrosis factor-alpha production, which in turn suppressed erythroid cell proliferation.

Some examples where in vivo whole animal bioassays have been used are: erythropoietin, where incorporation of radioactive iron into red blood cells is measured in exhypoxic polycythemic mice (Figure 10); human growth hormone, where daily weight gain of hypophysectomized female rats is measured; and human insulin, where the glucose level is measured 1 and 2.5 hours after injection in rabbits.

Insulin provides an excellent example of the evolution of the bioassay. Because there is now a better understanding of all aspects of insulin structure and function than ever before, from molecular identity and configuration to the physiological integration of biologic activity, useful assays may be more readily designed to address the various sides of a problem.

The recombinant human insulin now manufactured for treatment of diabetes may be screened for different aspects of molecular integrity affecting stability, pharmacokinetics, and therapeutic efficacy by HPLC and other methods of instrumental analysis, radioreceptor assays, and radioimmunoassays utilizing monoclonal antibodies to particular, defined regions of the hormone molecule, rather than an in vivo bioassay employing the whole animal.

## 2. Cell Culture Bioassay

In vitro cell culture bioassays have several advantages over in vivo whole animal bioassays. Theoretically, product binds to its receptor on a cell and induces a biological effect. There is an assumption that a measurable response at the cellular level relates to therapeutic activity and can be verified in the early phase of assay development by comparing with the animal bioassay.

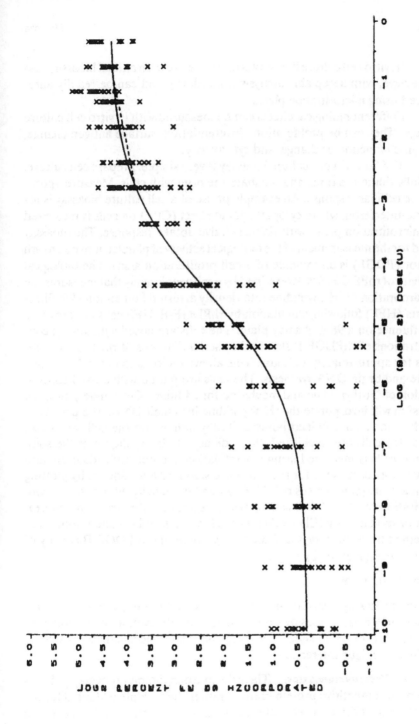

FIGURE 10   Exthypoxic Polycythemic Mouse Bioassay of Erythropoietin.

In vitro cell culture bioassay is more precise than animal bioassay, less expensive, with acceptable analysis time (days), and can be readily automated using microtitration plates.

Different biological effects can be measured with in vitro cell culture assays. They can be proliferation, differentiation, surface antigen change, enzyme induction or change, and cytotoxicity.

The flow of a cell culture bioassay is very simple: prepare cell culture, serially dilute the drug, and incubate the drug with cells. Measure appropriate cellular response. An example of use of a cell culture bioassay is for alpha-interferon, where cytopathogenic effect (CPE) on cells is measured in microtitration plates with viable cell dye uptake response. The bioassay used for the measurement of the biological activity of platelet derived growth factor (PDGF) is an example of a cell proliferation assay. The biological activity of rhPDGF-B is determined by a mitogenic assay that measures the incorporation of $^3$H-thymidine into density arrested human foreskin fibroblasts (HFF) following stimulation by rhPDGF-B. HFF cells are grown to confluence on a 96-well assay plate. The cells are dosed with known concentrations of rhPDGF-B standard or serial dilutions of rhPDGF-B samples for approximately 18 hours. This allows the cells to enter the mitotic cycle and begin DNA synthesis. The cells are pulsed with a fixed concentration of $^3$H-thymidine and incubated for 24 hours. Cells undergoing cell division will incorporate the $^3$H-thymidine into their DNA. The plates are washed to remove unincorporated $^3$H-thymidine and the cells are fixed. After the cells are solubilized, the radioactivity in an aliquot of the solubilized cells is measured using a scintillation counter. Individual dilution curves are constructed for the reference standard and samples by plotting radioactivity (cpm) versus rhPDGF-B standard activity dilutions. The concentrations of rhPDGF-B activity which result in half-maximal incorporation of radioactivity ($ED_{50}$ values) of reference standard and samples are obtained from the curves and used to calculate the rhPDGF-B activity of the samples (Figure 11).

## 3. Immunoassay

An immunoassay is any method that measures the concentration of a particular molecular species by exploiting the specificity of an immunological process. Most immunoassays are based on the physicochemical aspects of antibody–antigen interactions.

*a. Radioimmunoassay.* The principle of radioimmunoassay (RIA) is based on competitive protein binding rather than simple isotope dilution. A tracer quantity of a radiolabeled form of the antigen to be measured

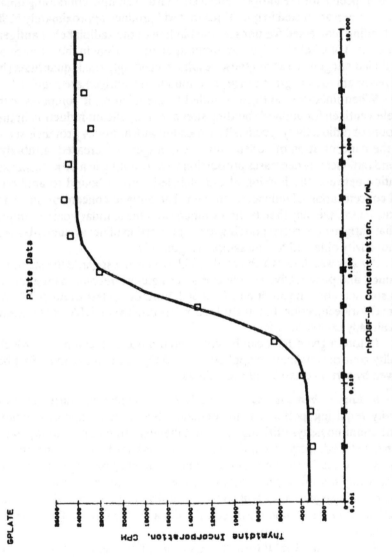

FIGURE 11    Mitogenic assay for the determination of the biological activity of rhPDGF-B.

(usually I-125 bound to the tyrosine in the case of proteins, and H-3 or C-14 substituents in the case of other antigens), an aliquot of highly diluted antiserum specific for the antigen, and a standard or sample containing unlabeled antigen are mixed in quantities that will produce approximately 50% net binding (corrected for nonspecific binding of the radiolabeled antigen to proteins) of labeled antigen by antibody at equilibrium in the absence of unlabeled antigen. To attain these results, exceedingly small quantities (in the picogram to nanogram range) of antibody and antigen are required.

When unlabeled antigen is added to the mixture, it competes with labeled antigen for antibody binding sites and a significant reduction in the amount of radioactivity specifically associated with antibody protein is seen. As the concentration of added unlabeled antigen is increased, antibody-bound radioactivity decreases proportionately, resulting in a dose-response relation between the fraction of radiolabeled antigen bound to antibody and concentration of unlabeled antigen. The antigen concentration of an unknown sample may then be determined from the standard curve, simply as that antigen dose corresponding to the proportion of net antibody-bound radioactivity obtained for the sample (Figure 12).

A study was done (Corbo et al., 1992) which investigated the stability, potency, and preservative effectiveness of single-use recombinant epoetin alfa solution after the addition of 1.0 or 1.5 mL of bacteriostatic 0.9% sodium chloride injection. In this study correlation between RIA and bioassay was already established.

Before a given RIA can be considered a useful technique to which quality control tests may be applied, the validity of the assay must first be proven by certain established procedures.

*b. Enzyme Immunoassay.* Since RIA may involve exposure to radioactivity, requiring special handling of materials and use of expensive equipment, immunoassays utilizing enzyme-labeled antigen or antibody have been developed. Enzyme immunoassay (EIA) is based on the same principles as RIA and its variants, but determines quantity of a ligand involved in the antigen-antibody interaction by measuring enzyme activity in the presence of a large excess of substrate. Typically, the product of the enzyme reaction exhibits a characteristic color and may be quantitated in a simple colorimeter.

A particularly useful form of EIA is enzyme-linked immunosorbent assay (ELISA), which detects and measures antibodies to a specific antigen. The method involves immobilization of the antigen on the surface of microtiter plate wells. The sample to be tested for antibody is then incubated in the wells. After washing, a second antibody covalently coupled to an

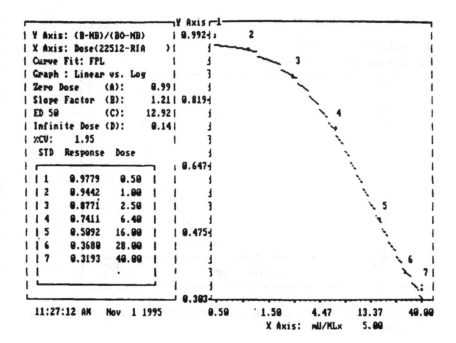

FIGURE 12  Radioimmunoassay for erythropoietin.

enzyme, usually horseradish peroxidase or alkaline phosphatase, is added. After another incubation and washing, a colorless substrate that is converted to a colored product by the enzyme is introduced. The reaction is stopped after an appropriate time, and is measured as the optical density of product at a characteristic wavelength. EIA can detect in the range of nanogram–microgram (Figure 13).

Immunoassays also assume that an antigen–antibody binding correlates with therapeutic activity. Often it cannot discriminate between biologically active versus biological inactive material. Immunoassays are inexpensive, with acceptable analysis time, and can be readily automated.

## 4. Biochemical Assay

The validity of biochemical assays was established during the early stages of biological investigation leading to the identification of a molecule. Most biochemical assays are dependent upon classical biochemical reactions established in the literature. The flow of these assays is very simple:

Curve Fit   4-Parameter                    Corr. Coeff:   0.999

$$y = (A-D)/(1 + (x/C)^{\wedge}B ) + D$$

A = 0.00497   B = 0.954   C = 248.   D = 6.68

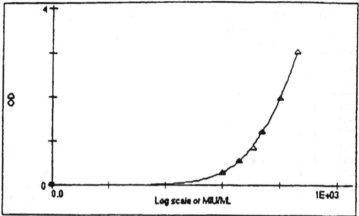

FIGURE 13   Enzyme immunoassay for the determination of the concentration of analytes.

prepare reagents, add drug and measure biochemical change by colorimetric methods.

These assays have very high precision and accuracy, they can mimic biological function, are suitable for a large number of samples, and can readily be automated. An example of a biochemical assay is the tissue plasminogen activator, currently marketed for treatment of acute myocardial infarction. Another example of a biochemical assay is the bioassay used for tissue factor. Tissue factor is a component of the extrinsic pathway for blood coagulations. Tissue factor is required for the stimulation of factor VIIa, a protease responsible for the conversion of factor X to factor Xa through proteolytic cleavage. Factor Xa, also a protease, can act to cleave the inactive factor VII to factor VIIa. Antibody to tissue factor blocks the activation of factor VIIa by preventing the interaction between tissue factor and factor VIIa.

In the assay for tissue factor, factor VII, factor X, tissue factor, and anti-tissue factor are combined in a buffer containing calcium. During the room temperature incubation, the presence of contaminating amounts of factors VIIa and Xa initiates a proteolytic cascade, leading to increasing

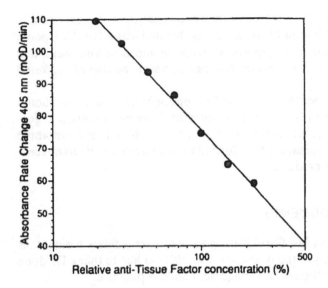

**FIGURE 14** Biochemical assay for the determination of the concentration of tissue factor.

amounts of factor VIIa and factor Xa. This cascade cannot occur in the absence of tissue factor or calcium. After a 15-minute incubation period the reaction is stopped by the addition of EDTA/EGTA. Increasing amounts of antitissue factor produce decreasing amounts of factor Xa. The proteolytic activity of factor Xa is measured colorimetrically in the absence of calcium using the substrate Spectrozyme fXa®. Increase in absorbance at 405 nm is measured in a microplate reader. The antibody used as a standard was given an arbitrary potency of 100% at a concentration of 5.6 ng/ml. The potency of a sample antibody diluted to 5.6 ng/ml is read off the standard curve directly as percentage of the standard (Figure 14).

## III. CONCLUDING REMARKS

Selection of an appropriate analytical technique is very important during the early phase, developmental phase and/or market phase. The cost of research, development and production of biotechnology products is raising some general public concerns. During the current health care crisis in the United States, the increased costs associated with biotechnology products needs to be revisited. In terms of selection of bioassay, an excellent example is the evolution of insulin bioassay now performed using HPLC. Upon a better understanding of all aspects of structure and function, useful and inexpensive assays may be readily designed to lower the cost of marketed product.

Suitable methods for stability-indicating and release functions should be identified in the early stage of development. If we pay attention to general considerations of the current environment, one should cut down where appropriate the cost of production, formulation, and release of pharmaceutical biotechnology products.

## IV. ACKNOWLEDGMENTS

I would like to thank Ms. Carol Keiling for her extraordinary typing skills and help in assembly of the typescript. I would also like to thank Dr. John Bontempo for his constant encouragement and his patience.

## V. RECOMMENDED READING

Andrews AT. Electrophoresis Theory, Techniques and Biochemical and Clinical Applications. Oxford: Clarendon Press, 1989.
Benacerraf B, Unanue ER. Textbook of Immunology. Baltimore: Williams & Wilkins, 1979.
Campbell AM. Monoclonal Antibody Technology. In: Burdon RH, Van Knippenberg PH, eds. Assay Techniques. New York: Elsevier, 1984, p 33.
Chard T. An introduction to radioimmunoassay and related techniques. In: Burdon RH, Van Knippenberg PH, eds. New York: Elsevier.
Haschemeyer RH, Haschemeyer AEV. Proteins—A Guide to Study by Physical and Chemical Methods. New York: John Wiley, 1973.
In: Hames BD, Rickwood D. ed. Gel Electrophoresis of Proteins—A Practical Approach. New York: IRL Press at Oxford University Press, 1990.
In: Heam MTW, Regnier FE, Wehr CT, ed. High-Performance Liquid Chromatography of Proteins and Peptides. New York: Academic Press, 1983.
Lehninger AL. Principles of Biochemistry. New York: Worth Publishers, 1982.
Light A. Protein Structure and Function. New Jersey: Prentice-Hall, 1974.
Segel IH. Enzyme Kinetics Behavior and Analysis of Rapid Equilibrium and Steady-State Enzyme Systems. Enzyme Assays. New York: John Wiley, 1975, p 77.
Stewart WE. The Interferon System. Interferon Assays. New York: Springer-Verlag, 1979, p 13.
Stryer L. Biochemistry. San Francisco: W. H. Freeman, 1975.

## REFERENCES

DePaolis AM, Advani JV, Sharma BG. Characterization of Erythropoietin Dimerization. Journal of Pharmaceutical Sciences 84:1280, 1995.

Trotta PP, Le HV, Sharma BG, Nagabhushan TL. Isolation and Purification of Human Alpha Interferon, a Recombinant DNA Protein. Developments in Industrial Microbiology 27:53, 1987.

Nagabhushan TL, Surprenant H, Le HV, Rosecki R, Levine A, Reichert P, Sharma B, Tsai H, Trotta P, Bauseh J, Foster C, Gruber S, Hoogerheide J, Mercorelli S. Characterization of Genetically Engineered Alpha-2 Interferon. In: Zoon KC, Noguchi PD, Liu TY, eds. Interferon: Research, Clinical Application and Regulatory Consideration. New York: Elsevier Science Publishing, 1983, p 79.

DePaolis A, Sharma B. Tryptic Map Variation of Erythropoietin Resulting from Carboxypeptidase B-Like Activity. Journal of Liquid Chromatography 17: 2777, 1994.

Kroon DJ, Baldwin-Ferro A, Lalan P. Identification of Sites of Degradation in a Therapeutic Monoclonal Antibody by Peptide Mapping. Pharmaceutical Research 9:1386, 1992.

Nagabhushan TL, Sharma B, Trotta PP. Application of Recycling Isoelectric Focusing for Purification of Recombinant Human Leukocyte Interferons. Electrophoresis 7:552, 1986.

Wu D, Regnier FE. Sodium Dodecyl Sulfate-Capillary Gel Electrophoresis of Proteins Using Non-cross-linked Polyacrylamide. J Chromatogr 608:349–56, 1992.

Ganzler K, Greve KS, Cohen AS, Karger BL, Guttman A, Cooke NC. High-performance Capillary electrophoresis of SDS-protein Complexes Using UV-transparent Polymer Networks. Anal Chem 64:2665–71, 1992.

Werner WE, Demorest DM, Stevens J, Wiktorowicz JE. Size-dependent Separation of Proteins Denatured in SDS by Capillary Electrophoresis Using a Replaceable Sieving Matrix. Anal Biochem 212:253–8, 1993.

Guttman A, Nolan J. Comparison of the Separation of Proteins by Sodium dodecyl Sulfate-slab Gel Electrophoresis and Capillary sodium Dodecyl Sulfate-gel electrophoresis. Anal Biochem 221:285–9, 1994.

Kroon DJ, Goltra S, Sharma B. Analysis of Monoclonal antibodies by Sodium Dodecyl Sulfate-capillary Gel Electrophoresis with Special Reference to Quantitation of Half Antibody. J Capillary Electrophoresis 2:1, 1995.

Udupa KB, Sharma BG. Possible Role of Tumor Necrosis Factor-Alpha in Erythropoietic Suppression by Endotoxin and Granulocyte/Macrophage Colony-Stimulating Factor. American Journal of Hematology 52:178, 1996.

Corbo DC, Suddith RL, Sharma B, Naso RB. Stability, Potency and Preservative Effectiveness of Epoetin Alfa after Addition of a Bacteriostatic Diluent. AJHP 49:1455, 1992.

# 7

## Membrane Filtration Technology

**FORREST BADMINGTON**
Millipore Corporation, Bedford, Massachusetts

## I. INTRODUCTION

This chapter deals with membrane filtration, a technology that has been in existence for many years. During this time, there have been many innovations and improvements and numerous books and technical papers have been published on the subject. In this chapter we will concentrate on the filtration applications specifically used during the processing and purification of biopharmaceutical products.

During World War II, the Germans began experimenting with membrane filters for evaluating the bacterial contamination of drinking water supplies. In the years following the war this technology was refined to the point where membrane filters became commercially available for a variety of filtration applications. In the 1960s, membrane filters were first introduced for sterilizing applications. In these applications, the filter membrane had to be extremely durable, integrity testable (before and after filtration) and capable of removing 100% of all microorganisms in the process solution. Today, pharmaceutical and biotechnical manufacturers rely on these filters to produce a sterile product that cannot be purified by other means.

Over the past several years, there have been a number of review articles (Weismantel, 1986; Johnston, 1992; Baccaro, 1993) written on filtration theory and techniques. There are also several excellent books written on the field of membrane filtration and separation technology

(Meltzer, 1987; Rousseau, 1987; Johnston, 1990; Winston Ho and Sirkar, 1992; Dickerson, 1992; Olson, 1995). In this chapter a review of the basic theory of filtration will be presented, but the main objective of the chapter is to assist you in determining what is the best type of filter train or system for your separation and then to optimize it for your specific process solution. It is intended to be used as a guide, with key references provided to give you the most recent detailed information in each specific area discussed. Other key resources available to help you with your filtration needs are independent consultants that specialize in the area of biopharmaceutical filtration applications and the technical filter specialists associated with a filter manufacturer. They have the knowledge and generally the experience necessary to assist you with your specific filtration needs.

Key sections in the chapter will include:

I. Introduction
II. Basic Filtration Theory
III. Filter Classifications and Characteristics
IV. Filter Performance Criteria
V. Filter Validation and Regulatory Issues
VI. Membranes and Filtration Systems
VII. Filter Selection and System Sizing
VIII. Filter Applications Specific to Protein Processing
References

## II. BASIC FILTRATION THEORY

### A. Particle Type and Size

The key to selecting an optimized filtration system is to understand the particle–media interactions with the filter matrix. In order to evaluate particle–media interactions, we need to know the type of particle involved, the capture mechanism that is occurring, and the type of filter media that is being challenged.

There are basically two particle types: (1) hard (nondeformable) particles, and (2) soft (deformable) particles. Typical hard particle contaminants that would be encountered in process solutions would include dust, sand, metal fines, and diatomaceous earth. Typical soft contaminants would include colloids, gels, bacteria, clays, and carbohydrates. The particles, contaminants, and microbiological organisms that are removed from various protein solutions with microporous membranes can vary in size from large 100-$\mu$m protein aggregates down to the smallest bacteria in the 0.2-$\mu$m range. The terms micrometers, microns, and $\mu$m, all referring to a size of

$10^{-6}$ m, are used interchangeably throughout the literature to refer to the size of the contaminants and microorganisms that are removed by membrane filters. Figure 1 includes the particle size of some common materials and the pore size distribution of typical membrane filters that would be used to remove or separate these materials from a solution.

## B. Particle Capture Mechanisms

The capture mechanisms operating in liquid and gases have been extensively studied (Rubow, 1981; Rubow and Liu, 1986; Rubow et al., 1987).

There are basically six capture mechanisms that occur in liquids and gases.

> *Size exclusion or sieving.*   Occurs when the particles are larger than the pores in the membrane.
> *Adsorption or electrostatic deposition.*   Occurs when there are attractive forces between the particles and the filter matrix.
> *Inertial impaction.*   Occurs when the inertia on the particles cause them to impact on the filter matrix. The inertia is developed as flow is diverted around the filter structure.
> *Interception.*   Occurs when particles in the flow stream contact the filter matrix.
> *Diffusion.*   Occurs when small particles move in random or Brownian motion to increase their probability for a collision with the filter matrix.
> *Gravitational settling.*   Occurs when large particles settle out in slow moving flow streams.

Depending on whether a gas or a liquid is being filtered, different capture mechanisms will predominate and directly influence the retention capabilities of the membrane filter. In gas filtration, the primary capture mechanisms are size exclusion, inertial impaction, diffusion, and electrostatic deposition. The capture mechanisms that predominate change as the particle size and the particle velocity changes. Small particles are retained by diffusion, larger particles by inertial forces. Once a particle is retained on the filter's surface, it is difficult for that particle to become dislodged due to the electrostatic and intermolecular forces applying over the small distances between the particle and the filter surface. The open cell-like structure of a defined membrane filter causes a particle to be exposed to a large surface, and to travel a very tortuous path through the filter. These effects magnify the retention mechanisms in gases resulting in increased

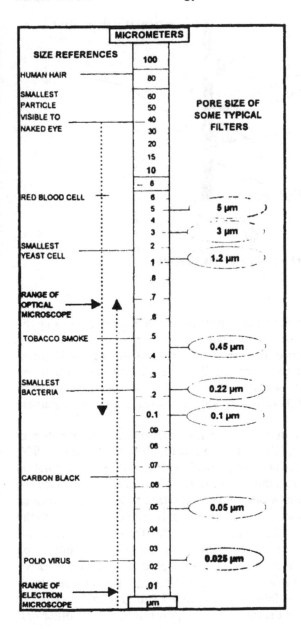

FIGURE 1   Micrometer ($\mu$m) sizes of common materials and the pore size of some typical membrane filters used to remove them (reprinted with permission from Millipore).

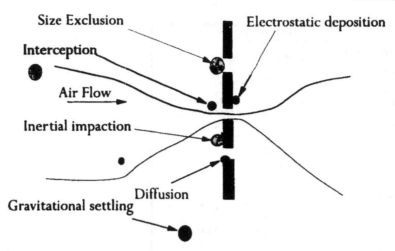

**FIGURE 2**  Retention mechanisms with gas streams.

particle retention. The six retention mechanisms that can occur during the filtration of gases are illustrated in Figure 2.

The primary capture mechanisms that occur during liquid filtrations are size exclusion and adsorption. During sterile filtration with membrane filters, the absolute retention of microbiological organisms occurs by size exclusion. Any effects of adsorption will increase the filter's particle retention.

In liquid filtration, there have been several theoretical models presented for filter plugging (Hermans and Bedee, 1935, 1936; Grace, 1956; Hermia, 1982) over time. The *caking* model describes filter cake formation through particle buildup on the surface of the filter, rather than inside the filter pores. *Gradual pore blocking* is characterized by gradual, controlled blocking of pores as a function of the amount of filtrate passing through the filter. Since particle contaminants are generally smaller than the pores, particles reaching the filter surface enter the pores and gradually build up on the inner walls until the pore is completely plugged. In *complete blocking*, a particle that reaches the filter surface completely blocks the pore. *Intermediate blocking* behavior falls between caking and gradual blocking, so particles may either block the inner walls of pores or adhere to other particles on the surface.

The type of filter plugging encountered depends on the type of contaminant particles in the process solution. Hard (nondeformable) particles generally produce a filter cake, while soft (deformable) particles often lead to complete or gradual pore blocking. Biological solutions contain mostly soft, deformable particles and typically block filters by gradual pore blocking or complete blocking.

## III. FILTER CARTRIDGE CLASSIFICATIONS AND CHARACTERISTICS

### A. Position in Filter Train

Filter cartridges can be classified by their position or function in the filter train (clarifying filter cartridge, prefilter cartridge, final filter cartridge) or by the structure of the cartridge (depth filter cartridge, surface filter cartridge, membrane filter cartridge).

Figure 3 shows how filter cartridges are classified by their position in the filter train. The clarifying filter cartridge would be the first filter cartridge in the train and the first filter to see process solution, the prefilter would be next and then the final filter. In this case, the clarifying filter is the most open filter in the train and would remove the largest particles from the process solution. Its main function is to extend the life of the prefilter. The prefilter in turn is used to extend the life of the final filter. Generally filter trains contain a prefilter and a final filter, however, in the filtration of extremely dirty solutions or solutions with a large number of particles, you may find a clarifying filter is also needed. In these filter applications, the clarifying filter may be used to protect several prefilters in series.

### B. Filter Structure

Filter cartridges are also characterized by the structure of the filter medium used in the cartridge. Using this classification, they would be referred to as depth, surface, or membrane filter cartridges.

The depth filter consists of fibrous, granular, or sintered materials that produce a random porous structure. When fluid passes through a depth filter, particles become trapped in its tortuous network of flow channels. The construction of depth filters varies. The filtration material may be wound (cotton, polypropylene), or may consist of resin-bonded laminates (most often, cellulose, acrylic, or glass fiber). Scanning electron micrographs in Figure 4 show the random porous structure of a glass fiber depth filter (a) in comparison to the regular porous geometric structure found in a sterilizing grade membrane filter (b). To add capacity to the depth filter, the thick filter matrix (3 to 20 mm) is constructed in a gradient density format. Gradient density means that the filter material becomes progressively more

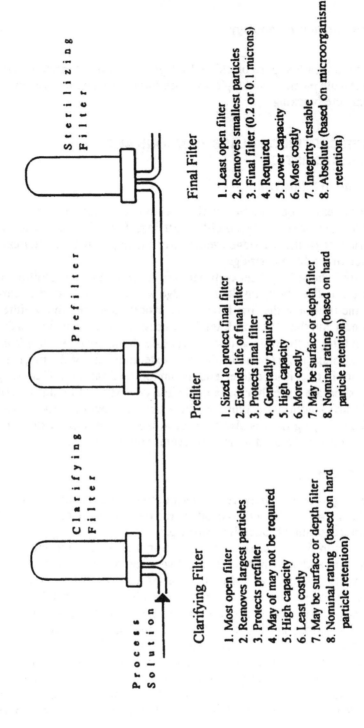

**Clarifying Filter**

1. Most open filter
2. Removes largest particles
3. Protects prefilter
4. May of may not be required
5. High capacity
6. Least costly
7. May be surface or depth filter
8. Nominal rating (based on hard particle retention)

**Prefilter**

1. Sized to protect final filter
2. Extends life of final filter
3. Protects final filter
4. Generally required
5. High capacity
6. More costly
7. May be surface or depth filter
8. Nominal rating (based on hard particle retention)

**Final Filter**

1. Least open filter
2. Removes smallest particles
3. Final filter (0.2 or 0.1 microns)
4. Required
5. Lower capacity
6. Most costly
7. Integrity testable
8. Absolute (based on microorganism retention)

**FIGURE 3** Filtration train containing a clarifying filter, prefilter, and final filter.

(a)

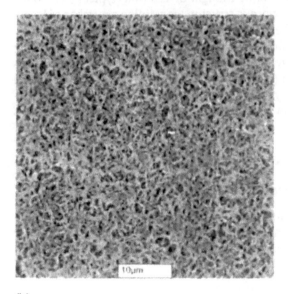

(b)

FIGURE 4   Scanning electron micrographs of a depth filter and a 0.2 μm-membrane filter. (a) shows the random porous structure of a glass fiber depth filter. (b) shows the regular porous geometric structure found in a 0.2 μm-membrane filter (reprinted with permission from Millipore).

retentive or tighter as you approach the cartridge core. The depth filter cartridge is generally used to remove the large particles (0.5 to 100 microns) in the process solution. Most filter manufacturers give depth filters nominal ratings, usually in the range of 90 to 99% retention efficiency of hard particles.

The surface filter is made up of multiple layers of media, usually glass, or polymeric microfibers that are cast on to a paper or polyester support layer. When a fluid passes through a surface filter, particles larger than the spaces within the filter matrix are retained, primarily on the surface. In addition, smaller particles tend to be trapped within the matrix, giving a surface filter properties of a depth filter as well. The surface filter usually has a higher retention efficiency rating than a depth filter and is used to remove the smaller particles (0.2 to 5.0 $\mu$m range) in the process solution that would tend to plug the final sterilizing grade membrane filter. Surface filters are given nominal retention ratings in the 99.0 to 99.99% retention efficiency range based on a hard particle challenge.

The membrane filter can be thought of as a thin (typically 0.1 to 0.2 mm) geometrically regular porous structure containing a number of layers of interconnecting cells. Sterilizing grade membrane filters are given absolute retention ratings. However, the absolute rating on a 0.2-$\mu$m sterilizing membrane is specifically related to the complete removal of microorganisms by the size exclusion (sieving) mechanism and not based on a hard particle retention efficiency test. The membrane filter and its unique characteristics are described in detail in the *Filtration Systems* section of this chapter.

## IV. FILTER PERFORMANCE CRITERIA

### A. Key Performance Factors Critical for Filter Selection

There are four key factors that are critical in evaluating the overall performance of a filter cartridge. They include (1) retention efficiency, (2) capacity (throughput), (3) differential pressure drop versus flow rate, and (4) compatibility and extractables.

### 1. Retention Efficiency

The retention efficiency of a filter is a measure of the ability to capture and retain particles. These particles are effectively removed from the solution passing through the filter. For a retention efficiency value to be useful it must include both a particle removal size and a percentage for the number of particles that will be removed at that size and larger. Filters are given both nominal and absolute ratings. Absolute in the classical definition implies 100% removal, and this rating is generally reserved for sterilizing grade membrane filters. In this case you can measure a 100% removal of all microorganisms.

Nominal ratings are given to filters that remove something less than 100% of the particles or contaminants at the filter rating. Unlike microorganisms, it is not possible to accurately measure the complete absence of very small particles or contaminants downstream of a filter. Filter manufacturers have not standardized on the use of the two terms, nominal or absolute, nor have they agreed on the percent removal given to a filter cartridge with a nominal rating. Some filter manufacturers will rate filters with 99.99% removal efficiencies as absolute, while others will give this filter a nominal rating. Therefore, it is important to have the removal efficiency stated as a percent.

## 2. Capacity (Throughput)

The dirt holding capacity (throughput) of a filter can be defined as:

1. The mass of contaminant held by the filter
2. The volume of fluid fed to the filter
3. The mass of contaminant fed to the filter

In practical applications, the capacity of a filter is a measure of the total volume of fluid that can be processed through a filter before it reaches a differential pressure across the filter that exceeds the limits of the filter system or when the flow through a filter is decreased to a rate that is unacceptable for the required processing conditions. In general, the capacity of a filter will vary dramatically depending on the type and particle size distribution of the contaminants being filtered. Therefore, estimates of filter capacities for specific filtration applications should only be done using a representative process solution.

## 3. Flow Rate versus Differential Pressure

One of the most important considerations when sizing a filtration system is the differential pressure drop that will be obtained across each filter. The maximum allowable pressure drop will control the minimum membrane area needed to do the job. The differential pressure across a filter is defined as the pressure upstream of the filter minus the pressure downstream of the filter at a specified flow rate. Filter manufacturers supply this data in the form of a pressure drop curve. The pressure drop curve shown in Figure 5 gives the clean water pressure drop for a 10-in. filter cartridge at various

## 0.2 micron Sterilizing Cartridges

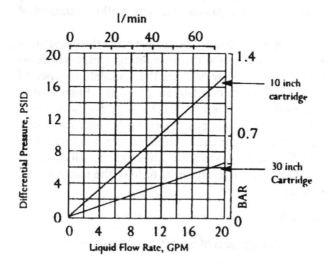

**FIGURE 5** Typical clean water pressure drop versus flow rate curves for microporous filter cartridges.

flow rates and shows that there is a linear relationship between pressure drop and flow rate over the acceptable flow range for this cartridge filter. For most cartridge filters, the clean water pressure drop across the cartridge increases linearly with flow rate, therefore, in many instances pressure drop data is listed as the differential pressure (psid) at a specified flow rate, generally one gallon per minute (gpm). In this example, the viscosity of the clean water solution was 1.0 centipoise (cp), however, a solution with a different viscosity will have a different pressure drop. A cold water solution, having a higher viscosity than a warm water solution, will exhibit a higher pressure drop on the same cartridge at equivalent flows. Generally for membrane filters, with solutions that are Newtonian, the pressure drop across the filter will increase linearly with an increase in viscosity. For example, a filter having a pressure drop of 0.5 psid in water (1.0 cp) would have a pressure drop of 5.0 psid in a solution with a viscosity of 10 cp. It is important to note that in the more open depth prefilter cartridges, viscosity may not be directly linear with pressure drop. In these applications, ask the filter manufacturer to supply viscosity versus pressure drop data for the specific filter cartridge in question.

Since water and most liquids are incompressible, the pressure differential across the filter will be independent of the absolute pressure of the system. This is not true for gases however, and, therefore, filter manufacturers provide pressure drop versus flow rate curves for gases at various pressures.

## 4. Compatibility and Extractables

Extractables are defined as the total amount of material that is extracted from a filter during the time it is in contact with the process solution. An ideal, fully compatible filter material would be nonreactive with the process solution. It would neither adsorb or remove any active ingredients from the process solution nor would it impart any extractables back into the solution. Filter manufacturers have developed an excellent array of inert polymeric membrane filters that are compatible with the many solutions requiring filtration in the biopharmaceutical industry. Though compatible with the process solution, these membrane filters may have a small quantity (milligrams) of material extracted over time. Regulatory agencies require and filter manufacturers supply extractables data for their specific filter products. Generally, the quantity of extractable material is so small that it cannot be detected in an actual process solution even with the most precise and sophisticated analytical equipment. What is generally provided by the filter manufacturer is the quantity and type of material extracted using a clean solvent that is representative of the solvent found in the process solution (Stone et al., 1994). The filter manufacturer should also specify the exact test conditions used for the extraction.

## B. Determining Filter Integrity

To insure that a filter, filter cartridge, or filtration system meets their critical process performance parameters, integrity tests have been developed to test and to verify filter retention efficiency. In the case of sterilizing grade filters, the critical performance criteria is the absolute (100%) retention of all microorganisms when tested under HIMA Guidelines.

To test the integrity of sterilizing grade filters either a nondestructive or a destructive test can be employed. Filter manufacturers perform the destructive test on a limited basis as lot release criteria, but, perform a nondestructive integrity test on each sterilizing grade filter to insure its integrity prior to shipment. The FDA Guidelines require the end-user of the filter to do a nondestructive integrity test on filters prior to their use in the processing of large volume parenterals (LVPS) and small volume parenterals

(SVPS). In addition, the FDA requires that integrity testing documentation be included with batch product records.

Destructive integrity testing involves challenging a filter with bacteria in accordance with HIMA methodology. Destructive challenge testing is the best way to determine if a filter or fabricated device meets the critical

**TABLE 1** Relationship of LRV Versus Water Bubble Point for Hydrophilic Durapore® (reprinted with permission from Millipore)

| Water bubble point (psi) | LRV[a] | Water bubble point (psi) | LRV[a] |
|---|---|---|---|
| 52.0 | >10[b] | 39.5 | >10.5 |
| 52.0 | >10 | 38.5 | > 9.2 |
| 52.0 | >10 | 38.0 | > 9.4 |
| 52.0 | >10 | 37.5 | >10.5 |
| 52.0 | >10 | 37.5 | > 9.4 |
| 50.0 | >10 | 37.5 | > 9.4 |
| 50.0 | >10 | 36.5 | > 9.2 |
| 50.0 | >10 | 36.0 | > 9.4 |
| 50.0 | >10 | 35.5 | > 9.4 |
| 50.0 | >10 | 35.5 | > 9.2 |
| 45.0 | >10 | 35.0 | > 9.4 |
| 45.0 | >10 | 34.5 | > 9.2 |
| 45.0 | >10 | 34.0 | 10.0 |
| 45.0 | >10 | 33.0 | 5.3 |
| 45.0 | >10 | 31.0 | 3.8 |
| 45.0 | > 9.5 | 31.0 | 3.7 |
| 45.0 | > 9.5 | 31.0 | 3.4 |
| 45.0 | > 9.5 | 29.0 | 3.3 |
| 44.0 | > 9.2 | 29.0 | 4.2 |
| 44.0 | > 9.5 | 28.0 | 3.9 |
| 44.0 | > 9.5 | 28.0 | 2.1 |
| 43.5 | >10.5 | 28.0 | 2.1 |
| 42.5 | >10.5 | 28.0 | 2.1 |
| 40.5 | >10.5 | 24.0 | 1.9 |
| 40.0 | >10.0 | 24.0 | 1.9 |
| 40.0 | > 9.4 | 20.0 | 0.5 |
| 40.0 | > 9.4 | 20.0 | 0.4 |

[a] $LRV = \log_{10} \dfrac{\text{organisms in challenge}}{\text{organisms in effluent}}$

[b] "Greater than" signs indicate sterile effluent.

performance criteria of a sterilizing filter. Nondestructive integrity testing may be done on filters pre- and postuse. Integrity testing of sterilizing filters prior to batch processing prevents using a nonintegral filter at the beginning of the filtration process. Integrity testing sterilizing filters after a batch has been completed can detect if the integrity of the filter had been compromised during the filtration process.

There are two nondestructive tests routinely used in the biopharmaceutical industry to determine filter integrity, (1) the bubble point test, and (2) the diffusion test. The pressure hold, forward flow, and pressure decay tests are also referred to as integrity tests, but are variations of the diffusion test. The theory and practical application of these integrity tests have been reviewed extensively in the literature (Emory, 1989a, 1989b; Meltzer, 1989a, 1989b). To be able to use a nondestructive integrity test, physical tests were developed that correlate to the destructive bacterial challenge test. A specification for the physical test directly correlates to the bacterial challenge data. Once this correlation is established, it can be determined that a filter or filter cartridges passing the physical test is an integral sterilizing filter. An example of published data that correlates membrane bubble point to bacterial (*Brevundimonas diminuta*) retention is shown in Table 1. A sterilizing grade 0.2-$\mu$m Durapore$^R$ membrane will typically have a bubble point of 52psi with a minimum specification of greater than 45psi. The data shown in Table 1 shows there is a 10-psi bubble point margin built into the Durapore specification. Figure 6 illustrates how an air/water diffusion (cc/minute)

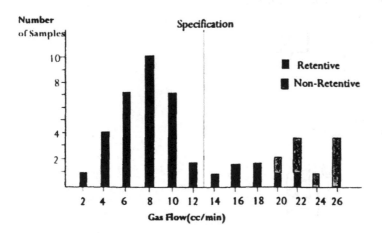

**FIGURE 6**  Air/water diffusion flow rates for retentive versus nonretentive sterilizing cartridges.

specification would be generated for a sterilizing grade filter cartridge. The diffusion specification for the sterilizing cartridge is set at a point well below the diffusion rate of nonretentive cartridges.

## 1. Bubble Point Test

The most widely used nondestructive integrity test is the bubble point test. Bubble point is based on the fact that liquid is held in the pores of the filter by surface tension and capillary forces. The minimum pressure required to force liquid out of the pores is a measure of the pore diameter.

Bubble point formula:
$$P = \frac{4k \cos \theta \sigma}{d}$$

where
$P$ = bubble point pressure
$d$ = pore diameter
$k$ = shape correction factor
$\theta$ = liquid–solid contact angle
$\sigma$ = surface tension

An in-process bubble point test will detect damaged membranes, ineffective seals, system leaks, and distinguish filter pore sizes. A schematic showing a typical bubble point test is listed in Figure 7. The test is performed by:

Bubble   Point   Test

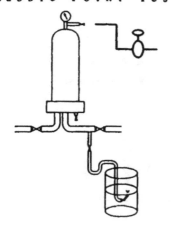

FIGURE 7   Set-up for a bubble point test on a single cartridge.

1. Wetting the filter with an appropriate fluid, typically water for hy-drophilic membranes or an alcohol/water mixture for hydrophobic membranes.
2. Pressurizing the system to at least 80% of the expected bubble-point pressure which is listed in the manufacturer's literature.
3. Increasing the pressure slowly until a rapid flow of continuous bub-bling is observed at the outlet of the tubing under water.

A bubble point value lower than the specification would indicate a nonintegral filter, however, the following could also lower the bubble point value:

Fluid with a surface tension different from the recommended test fluid remains in the filter.
The wrong pore size cartridge.
Cartridge at a high temperature.
A membrane filter not fully wet.

As seen from the bubble point equation, surface tension plays a sig-nificant role in the value that will be obtained. Therefore, if the surface tension of the solution you are filtering is dramatically different from water, it must be flushed out thoroughly from the filter cartridge prior to running the postfiltration bubble point integrity test.

## 2. Diffusion Test

At a differential gas pressure below the bubble point, gas molecules migrate through the water-filled pores of a wetted membrane filter following Fick's law of diffusion. The gas diffusional flow rate for a filter is proportional to the differential pressure and the total surface area of the filter. The diffu-sion test is run at approximately 80% of the minimum bubble point. The gas which diffuses through the membrane is measured to determine a fil-ter's integrity. The flow of gas is very small in low area filter devices, but it is significant in large area filters. Filter manufacturer's provide maximum diffusional flow specifications for their specific membranes and devices. Based on the diffusional flow rate, one can predict bacterial retention test results and, therefore, the integrity of the filter device. The schematic listed in Figure 8 illustrates a typical diffusion test. The test is performed by:

1. Wetting the filter with an appropriate fluid, typically water for hy-drophilic membranes or an alcohol/water mixture for hydrophobic membranes.

Diffusion Test

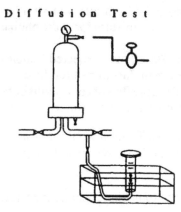

FIGURE 8   Set-up for a diffusion test on a single cartridge.

2. Slowly increasing pressure on the upstream side of the filter to the recommended test pressure provided by the manufacturer. Typically this pressure will be at least 80% of the bubble point.
3. Allow the system to equilibrate.
4. Measure the gas flow rate at the outlet for one minute with an inverted graduated cylinder or a flow meter.

A diffusional flow rate higher than the specification is an indication of a non-integral filter, however any of the following items could also increase the gas flow rate:

Cartridge at a high temperature
Incompletely wetted membrane
Inadequate stabilization time
Liquid/gas combination different from the recommended fluids

The pressure hold test, also known as a pressure decay or pressure drop test, is a variation of the diffusion test. In this test, a highly accurate gauge is used to monitor upstream and downstream changes due to gas diffusion through the filter. Because there is no need to measure the flow of gas downstream of the filter, you can eliminate any risk to downstream filter contamination. The pressure hold value is dependent on the diffusional flow and upstream volume. It can be calculated using the pressure hold equation shown below:

Pressure hold: $\dfrac{D(T)(P_a)}{V_h} = \Delta P$

where $D$ = diffusion rate (cc/minute)
$T$ = time (min)
$P_a$ = atmosphere pressure (14.7 psig)
$V_h$ = upstream volume of apparatus (cc)
$\Delta P$ = pressure drop (psig)

The pressure hold test is the general method employed by most automated integrity test units. The automated units can also determine bubble point by measuring changes in upstream pressure. The new automated integrity testers that are available are accurate, reliable, and computer controlled. They offer many advantages over a manual test in that they ensure consistency by removing operator subjectivity and error from the test procedure. The unit can be validated and will supply a permanent printed record of the integrity test results. An automated integrity test instrument with the capabilities described above is shown in Figure 9.

There has been confusion and inconsistency in the industry as to the correct type of nondestructive integrity test to use for large area filter devices. In small area filters ($< 1.0$ ft$^2$), where the diffusional flow rate is generally too low to measure accurately with available laboratory equipment, the bubble point test is usually recommended, either manually or with an automatic integrity test instrument. However, when the filter area is increased, especially beyond that of a 10-in. cartridge (approximately 7 ft$^2$), it is difficult to distinguish the increase in flow rate that will occur at the filter bubble point over the background diffusional flow through the filter. Therefore, in large area devices, a diffusion test is recommended. Generally the diffusion test is run at a single pressure. The controversy relates to the accuracy of a diffusion test run at one specific pressure, also referred to as a single point diffusion test. There have been several papers published (Emory, 1989a, 1989b; Meltzer, 1989a, 1989b, 1992) related to this issue. It is generally agreed that a combination of the bubble point and diffusion test, will improve the sensitivity of either test run alone. Basically, the combination bubble point and diffusion test requires the measurement of several gas diffusion rates at pressures below and up to the filter's bubble point. This procedure can generally be programmed into the new automatic integrity test instruments being sold on the market today. To determine the best integrity test for a specific filter device, look to the filter manufacturer for guidance. They should supply specific integrity test procedures for their sterilizing grade membrane filter cartridges and devices.

FIGURE 9    Automatic integrity test instrument (reprinted with permission from Millipore).

## C. Filter Adsorption and Shear Effects

### 1. Protein Binding

When filters are used to process biotechnology products containing proteins that are extremely expensive or at a level where any loss is considered detrimental to the overall quality of the product, it is important to consider and evaluate what effects the membrane materials will have on the process solution. Because most polymeric microporous membrane filters have a large internal area accessible to proteins there will be some nonspecific binding of proteins by the membrane. Certain polymer materials bind protein at higher rates than other materials. Certainly, hydrophobic polymeric membrane materials will have higher binding levels than those with hydrophilic surfaces. Filter manufacturers, therefore, provide modified hydrophilic membranes specifically for protein applications. Millipore Corporation has a modified PVDF membrane (GV) and Pall Corporation has a modified polyamide (BioInert) membrane for protein filtrations. Since the level of binding is going to be dependent on the type of protein or peptide and its concentration in solution, it is recommended that an evaluation be done with a specific process solution and several filter materials before finalizing on one filter type. It is important to evaluate the entire filter train including any clarifying or any prefilters that the process solution will be in contact with during the filtration cycle. There have been a number of papers published in this area in recent years (Pitt, 1987; Martin and Manteuffel, 1988; Van den Oetelaar et al., 1989; Sarry and Sucker, 1992a, 1992b; Datar et al., 1992; Hu et al., 1993; Nema and Avis, 1993). Table 2 (Pitt, 1987) lists data on the nonspecific binding of bovine serum albumin (BSA) and sheep immunoglobulin G (IgG) with several commercial polymeric membranes. In addition to protein binding, it is also important to monitor the adsorption of any other products or additives (preservatives) that may be in your process solution at low concentrations (Brose and Henricksen, 1994).

## 2. Shear Effects and Protein Degradation

In general, protein degradation, denaturation, or change in confirmation is due more to the interaction of the protein (adsorption) with a specific polymer filter material than to the shear effects that are created as the protein solution is passed through a microporous membrane (Truskey et al., 1989). However, in one experimental study (Bowen and Gan, 1992) on the interactions of the enzyme yeast alcohol dehydrogenase (YADH) with microfiltration membranes, a loss in enzyme activity was attributed more to the shear effects occurring during the filtration than to a filter material interaction. Since many filtration applications require the use of pumps, these

**TABLE 2**  Protein Adsorption to Polymer Surfaces at a 1000 milligrams per liter (mg/l) Protein Concentration

| Membrane type | $\mu$g BSA bound/cm$^2$ | $\mu$g IgG bound/cm$^2$ |
|---|---|---|
| Hydrophilic polyvinylidene fluoride membrane (PVDF) | 3.4 | 3.8 |
| Cellulose diacetate membrane (CDA) | 1.1 | 8.6 |
| Mixed esters of cellulose acetate and cellulose nitrate | 274 | 238 |

Data from Pitt, 1987 with permission.

devices should also be evaluated as a potential source that could contribute to shearing effects.

During the separation of mammalian cells from a protein product, shear effects may adversely affect the quality of the product passing through the filter. High flow rates and high differential pressures across the filter may cause cell lysis to occur. This increase in cell debris may plug the downstream filter or contaminate the final product. It is recommended that these separations be run at low flow rates and low differential pressures and that the amount of cell lysis occurring during the filtration be continually monitored.

## V.  FILTER VALIDATION AND REGULATORY AGENCIES

### A.  Filter Validation

In the 1987 "Guidelines on Sterile Drug Products Produced by Aseptic Processing," the U.S. Food and Drug Administration defines validation as "Establishing documented evidence which provides a high degree of assurance that a specific process will consistently produce a product meeting its predetermined specifications and quality attributes." Ultimately the drug manufacturer is responsible for final filter validation in its manufacturing process. They will, however, rely on the filter manufacturer to provide the necessary information and services to assist with the validation. It is important to chose a filter supplier familiar with all the regulatory agencies, and their current requirements. A list showing specific validation information that should be supplied by filter manufacturer's for their products is summarized in Table 3.

### B.  Regulatory Agencies

In the United States there are several key agencies involved in regulating aseptic filtration processing and providing requirements and test recommendations for validating filtration systems. They include the Food and Drug Administration (US Dept. of Health and Human Services, 1987), the Health Industry Manufacturers Association (HIMA, 1982), the U.S. Pharmacopoeia (USP, 1995), and the American Society of Testing Materials (ASTM).

In the European Union, the use and validation of aseptic processing via filtration is contained in "The Rules Governing Medicinal Products in the European Community," specifically Volume IV Annex 1 Manufacture of Sterile Medicinal Products (Commission of the European Communities,

TABLE 3  Validation Responsibilities of Filter Supplier and Filter User

| Parameter | Filter supplier | Filter user |
|---|---|---|
| Filter quality | Document all filter claims and manufacturing process | Review data and audit manufacturer |
| Filter consistency | Provide lot release data and document manufacturing process | Show drug product consistency |
| Sterilization procedure | Recommend sterilizing procedures with temperature and cycle time limits | Operate within manufacturers limits<br>Validate for their process application |
| Structural integrity | Provide filter usage limits | Validate for their process |
| Integrity testing | Provide specifications, test procedures and correlation data to bacterial retention | Validate for their process and solution |
| Fibers | Meet non-fiber releasing claim [21 CFR 210.(3b)] | Documentation for claim |
| Sterilizing grade filter | Provide information on release criteria | Validate with their process solution |
| Extractables | Provide extractables data type and amount, pass USP oxidizables test | Evaluate with their system and process solution |
| Compatibility | Provide compatibility tables | Evaluate with their solution |
| Toxicity | Provide test data showing product meets USP class VI plastics test and mouse safety test | Document and verify information |
| Adsorption | Provide known information | Evaluate with their solution |
| Product activity/ stability | Provide known information | Evaluate with their solution |
| Endotoxins | Provide lot release data | Evaluate filter and system |
| Particles | Provide particle retention data | Product meets USP particulate standards for injectables |

1992). However, it is important to note that the European Union is made up of many countries each with their own government agencies regulating and providing recommendations for aseptic processing via sterile filtration.

There are various Good Manufacturing Practices (GMP) regulations in Japan concerning the manufacturing and use of sterile filters. The major regulation is included in the "Manufacturing Control and Manufacturing Hygiene Control of Sterile Products etc. (PAB/IGD, Notification 119, October 9, 1980)."

In Australia the manufacturing of therapeutic goods falls under the "GMP Auditing and Licensing Section, Compliance Branch Therapeutic Goods Administration" (Australian Dept. of Community Services and Health, 1992).

The Health Industry Manufacturers Association (HIMA), in its Guideline (HIMA, 1982) defines the term "sterilizing filter" and discusses the methods that are to be used for testing a sterilizing grade filter. The HIMA guideline specifies that *Pseudomonas diminuta* (American Type Culture Collection 19146) be used as the challenge organism at a concentration of $10^7$ organisms per square centimeter of effective surface area (EFT). *Pseudomonas diminuta* was shown to be misclassified at the genus level (Segers et al., 1994) and has been recently reclassified into the genus *Brevundimonas* and renamed *Brevundimonas diminuta*. The new name will be used throughout this chapter.

The Food and Drug Administration (FDA) in their "Guideline on Sterile Drug Products Produced by Aseptic Processing" (US Dept. of Health and Human Services, 1987) defines a "sterilizing grade filter" as one which produces sterile effluent when challenged with *Brevundimonas diminuta* (*B. diminuta*) at a concentration of $10^7$ organisms per square centimeter of surface area. The choice of *B. diminuta* as the challenge organism was purposeful in that its size, shape, and degree of aggregation was a careful consideration. It is a small organism, with a size, shape, and degree of aggregation that can be consistently controlled in the laboratory when grown in saline lactate broth (SLB). The morphology of *B. diminuta* was determined by scanning electron microscopy image analysis and when grown in SLB had an average diameter of 0.32 μm (Leahy, 1983). The FDA Guidelines require integrity testing of filters used in the process of large volume parenterals (LVPS) and small volume parenterals (SVPS). In addition the FDA requires that integrity testing documentation be included with batch product records.

The FDA recommends challenge testing sterile filters under worst case processing conditions. It also recognizes that bacterial challenge test-

TABLE 4 USP 23 Tests for Sterile Products Used in the Processing of Pharmaceutical Solutions

| USP 23 test | Test description | Measurement criteria |
|---|---|---|
| <88> | Biological reactivity tests In vivo class VI plastics test | Extract from material when injected will not cause death or related systems in mice |
| <88> | Biological reactivity tests Mouse safety test | Extract from filter device will not cause death of related systems in mice |
| <788> | Particulate matter in injections | Filter device releases particles below current standard for process solution |
| <71> | Sterility tests | Sterile effluent from filter device |
| <85> | Bacterial endotoxins | Effluent from filter device below current endotoxin standard |
| <151> | Pyrogen test | Will not cause a pyrogenic response in a rabbit or extract from filter is below current LAL standard |
| Water for injection pp, 1635, 1636 | Oxidizables test | Passes the oxidizables substance test after a described flush period |

ing in all solutions is not practical, because in certain solutions and under certain test conditions, the organisms would not survive. For instance (Levy, 1987) was unable to challenge test filters with *B. diminuta* in the presence of $AlCl_3$ $Mg$ $Cl_2$ or Triton x-100 when concentrations approached 0.1 M.

The United States Pharmacopeia (USP 23, 1995) issues a series of tests and requirements that should be met by all sterilizing grade membrane filter devices before they can be used to process solutions or drugs in the pharmaceutical industry. Depending on the specific test it can be required as a one time test, on an audit basis, or in limited testing for lot release. A summary of the USP tests are listed in Table 4.

The American Society for Testing Materials (ASTM, 1988) has issued a variety of methods that are applicable to the evaluation and characterization of membrane filters. Specifically, test method F838-83 describes a test procedure for determining the bacterial retention of membrane filters used in liquid filtration applications.

## VI. MEMBRANES AND FILTRATION SYSTEMS

### A. Membranes

Generally applications using membrane filters will be classified as microfiltration (MF) or ultrafiltration (UF) depending on the size of the contaminants that are being removed. Microfiltration is the process of removing contaminants in the 0.025- to 5.0-$\mu$m range from fluids by passage through a microporous membrane filter medium. Ultrafiltration is the process of separating extremely small particles ($<0.05$ $\mu$m) and dissolved molecules from fluids. The primary basis for ultrafiltration separation is molecular size although secondary factors such as molecule shape and its charge can play a role. Materials ranging in size from 1,000- to 1,000,000-molecular weight are retained by ultrafilter membranes while salts and water will pass through. Ultrafilters are used to purify and collect material passing through the filter and material retained by the filter. Although nonmembrane or depth filter materials such as those found in fibrous media are used to separate micron-sized particles, only a membrane filter with its defined pore size and structure can insure a quantitative removal efficiency. Membrane filters may be used for final filtration or prefiltration, however the fibrous depth filter is generally only suitable for clarification and prefiltration applications.

The most widely used MF membranes are homogenous, but there are some skinned-asymmetric or anisotropic membranes on the market. Ultrafiltration membranes are mostly asymmetric and cover the pore size range of 0.001 to 0.05 $\mu$m.

Most membrane filters are produced from low-water-absorptivity polymeric materials, which are relatively hydrophobic. A few membranes, however, are being made from inorganic materials, mostly ceramics, and as technology increases in this area so will the number of membranes. Organic membranes are manufactured by several distinct processes which include phase inversion, sintering and stretching, track etching, and thin film composites. A group of scanning electron micrographs for a variety of flat stock membrane filters used for deadend applications are illustrated in Figure 10. Membranes are manufactured with flat sheet, hollow fiber, or tubular geometry. Most membranes are available in a flat sheet configuration for small scale filter applications or system sizing, but are generally fabricated into a variety of devices with various configurations and surface area for

(a)

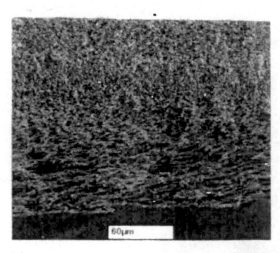

(b)

FIGURE 10 Scanning electron micrographs of various membrane filters: (a) symmetric membrane, (b) asymmetric membrane, (c) track etched membrane, (d) composite membrane (reprinted with permission from Millipore).

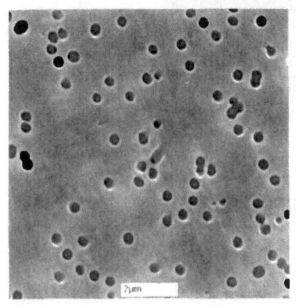

(c)

(d)

FIGURE 10 Continued

large scale filter applications. An extensive list of commercially available microfiltration and ultrafiltration membranes is included in Table 34-1 of the "Membrane Handbook" (Ho and Sirkar, 1992).

## B. Membrane Filter Operation

In actual filtration processes membranes can be run in a deadend operation or in a tangential flow filtration (TFF) operation. In a deadend operation, all the fluid is passed through the membrane filter and over time the retained particles gradually block the pores in the membrane or form a cake on the surface of the membrane to stop flow and plug the filter. In tangential flow, also referred to as cross flow, the bulk of solution is made to flow tangential or across the filter membrane. It has an advantage over deadend filtration in that the particles or contaminants on the surface of the membrane can be washed away for continuous long term membrane use. Usually ultrafiltration membranes, having a very small pore size, are run in the tangential flow mode and microporous membranes are run in the deadend mode. However, microporous membranes can be run in the tangential flow mode and their usage in this mode is on the increase. Microporous membranes run in the tangential flow mode have the ability to purify and concentrate particulate materials, such as bacteria, yeast, and animal cells where large molecules must pass through the filter. One of the main problems with running microporous membranes in the tangential flow mode is the control of concentration polarization or fouling on the surface of the membrane. Research is continuing in this area (Baccaro, 1993) and new membranes and techniques are being developed to reduce this concentration polarization phenomena.

## C. Membrane Filter Devices (dead end)

In order to perform the filtration with a flat stock sterilizing grade membrane, it must be placed into a suitable filter holder and properly sealed to prevent downstream contamination. Filter manufacturer's supply 47-, 90-, and 293-millimeter (mm) diameter stainless steel filter disk holders, specifically designed for this purpose (Fig. 11). The filter holder is equipped with a downstream membrane support screen, an O-ring (generally silicone) for sealing, and an upstream vent valve to allow trapped air to be purged from the system. The entire disk holder with membrane installed can be steamed or autoclaved and integrity tested prior to the filtration step. Specific procedures and instructions for using the various types of filter disk holders that are available are supplied by the manufacturer. Multi-disk holders

FIGURE 11    A 90-mm stainless steel filter disk holder (reprinted with permission from Millipore).

are also available and were used quite often in the past to process a large volume of solution. However, with the work required to clean, load, and sterilize the multi-disk holder and with the development of sterilizing grade pleated filter cartridges and stacked disk cartridges, multi-disk holders are rarely employed in new applications that require the sterile processing of bulk pharmaceutical solutions. A 293-millimeter sterilizing filter disk and stainless steel multi-disk holder is shown in Figure 12.

The most common sterilizing grade membrane filter element in use today for the sterile processing of large volumes of solution is the pleated cartridge illustrated in Figure 13. In this configuration, a membrane is pleated (folded) and wrapped around an open inner plastic core. Usually incorporated into the cartridge with the pleated membrane is an upstream and downstream support screen. The support screen holds the membrane in form during cartridge manufacture and provides added support to the membrane during cartridge use, especially where there may be a back pressure or back flow placed on the cartridge. The pleated filter pack and inner core are then sealed into two end caps, one acts as a plug, the second

**FIGURE 12**  A 293-mm sterilizing disk filter and stainless steel multi-disk filter holder (reprinted with permission from Millipore).

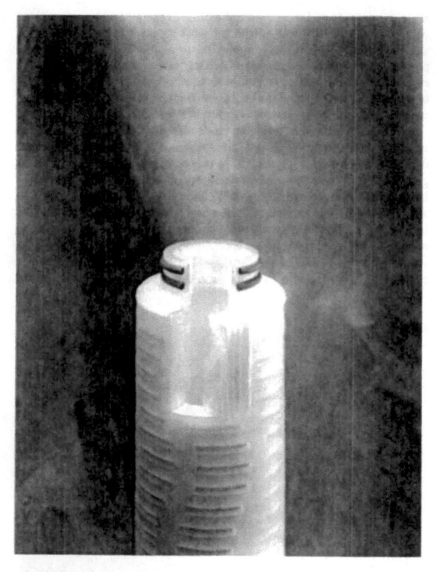

FIGURE 13  Pleated membrane cartridge filter (reprinted with permission from Millipore).

provides the exit flow opening usually equipped with a double O-ring seal. During use, the cartridges are placed into a stainless steel housing, which allow the cartridges to be steam sterilized and integrity tested prior to use.

There is usually between 4 to 8 square feet of membrane filter in a 10-in. pleated cartridge, depending on the membrane thickness, the pleat height, and the type of support material used. The exact area, referred to as effective filtration area (EFA) should be included in the manufacturer's specifications. Using the effective filtration area in a pleated cartridge device in combination with the data that is developed from running small scale plugging trials on flat stock disk material (47mm disks), it is possible to estimate the volume that will plug a 10-in. cartridge. This plugging test on flat stock membranes is known as a flow decay test and is discussed in detail in the section on filter selection and system sizing. Based on the disk data, it is recommended that a small pilot scale run be performed with small cartridges 2 to 4 inches in length. Using the pilot run data it is then possible to determine the approximate number of cartridges that will be needed for a large scale production run.

Cartridges are usually available in 2- to 40-in. lengths, however, the 10-, 20- and 30-in. lengths are the most commonly used. A series of sterilizing grade pleated cartridges from 10 inches in length up to 40 inches in length are shown in Figure 14. To process small to medium volumes of solution, the filter cartridge is placed into individual stainless steel housings as illustrated in Figure 15. For processing very large volumes (>5000 l.), large multi-round stainless housings such as those shown in Figure 16 are employed. For the sterile processing of very small volumes of pharmaceutical solutions there are many disposable filter cartridges and capsules available (Figure 17). These devices can contain either stacked membrane disks or pleated membrane material.

## D. Membrane Filter Devices (tangential flow)

Manufacturers offer a wide variety of tangential flow filtration (TFF) systems containing both microporous or ultrafiltration membranes. Systems are available for processing solutions from the laboratory bench top sample preparation scale (milliliters to liters) to large-scale production quantities (several thousand liters or larger). The Standard TFF systems available today are designed for sanitary and validatable operation. In many instances filter manufacturers will custom design a TFF system for your specific process application. Pictured in Figure 18a is a bench top TFF system capable of processing solutions from 10- to 500-l. and a TFF system (Figure 18b) used for serum protein concentration in the 1000 l. or greater range.

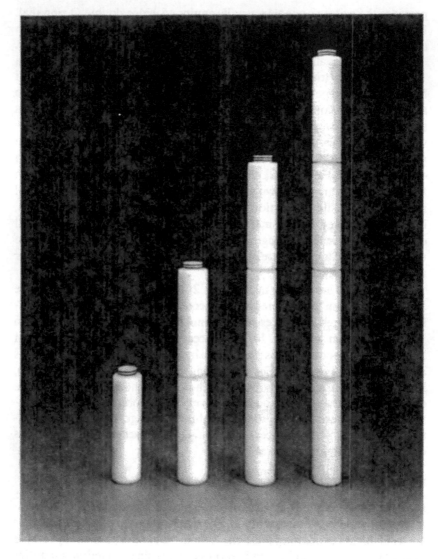

FIGURE 14   Pleated membrane filter cartridges from 10 inches to 40 inches in length (reprinted with permission from Millipore).

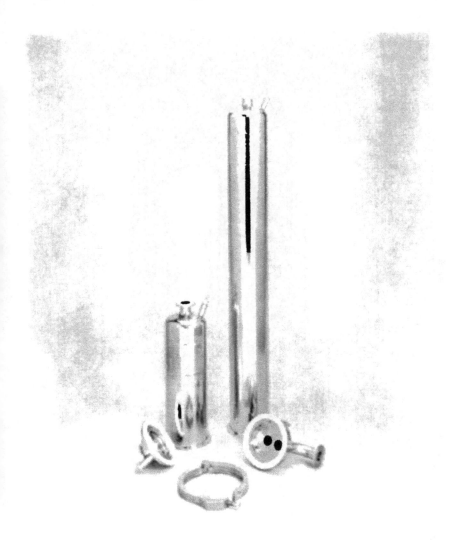

FIGURE 15 Stainless steel housings for single filter cartridges (reprinted with permission from Millipore).

(a)

FIGURE 16   (a) Large multi-round stainless steel filter cartridge housings, and
(b) multi-round housing with filters (reprinted with permission from Millipore).

(b)

FIGURE 17   Stacked membrane filter disk cartridges and capsules (reprinted with permission from Millipore).

### E. Filter Suppliers

There are a large number of manufacturers and suppliers of clarification and prefiltration filter cartridges, housings and accessories. However, due to the stringent regulatory requirements on the manufacture and release of sterilizing grade filters and filter cartridges, there are fewer manufacturers in this area. A partial list of sterilizing filter manufacturers is included in Table 5. Millipore Corporation and Pall Corporation are the two largest suppliers of sterilizing membrane filters and products on a world wide basis with a total combined annual 1996 sales of filtration products in excess of 1.5 billion dollars (Millipore, 1996; Pall, 1996).

(a)

**FIGURE 18**  Tangential flow systems: (a) bench top system capable of processing 10 to 500 liters, (b) large process system for serum protein concentration in the 1000's of liters range (reprinted with permission from Millipore).

## VII. FILTER SELECTION AND SYSTEM SIZING

### A. Liquids

1. Filter Selection

Filtration systems for aseptic processing in liquid applications generally consist of a series of filter cartridges, with each cartridge in the series protecting and extending the life of the next filter cartridge.

A typical filtration train used for processing human or mammalian serum is illustrated in Figure 19. Raw serum being processed through this train would become a sterile serum product. In the filter train, a series of

(b)

FIGURE 18   Continued

**TABLE 5**   Manufacturers of Sterilizing Grade Membrane Filters and Devices

| | |
|---|---|
| Domnick Hunter Filters Limited | Nucleopore Corporation |
| Durham Road, Birtley, Co | 7035 Commerce Circle |
| Durham, England DH3 2SF | Pleasanton, California 94566 |
| Tel: (091) 410 7621 | Tel: (415) 463-2530 |
| Millipore Corporation | Pall Corporation |
| 80 Ashby Road | 2200 Northern Boulevard |
| Bedford, Massachusetts | East Hills, New York 11548-1289 |
| 01730-2271 | Tel: (516) 484-5400 |
| Tel: (617) 275-9200 | |

Sartorius GmbH
Postfach 3243, Weender
Landstrasse 94-108, D3400
Goettingen, Germany
Tel: (0551) 308-492

clarification and prefiltration cartridges are used to protect and extend the life of the final filters. Pressure gauges are placed both upstream and downstream of each housing to monitor the pressure increase across each filter cartridge. Valves are strategically located to vent all housings and ports are placed on the sterilizing housings for integrity testing the filters inside. In

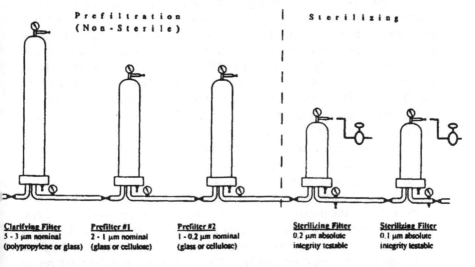

| **Clarifying Filter** | **Prefilter #1** | **Prefilter #2** | **Sterilizing Filter** | **Sterilizing Filter** |
|---|---|---|---|---|
| 5 - 3 μm nominal | 2 - 1 μm nominal | 1 - 0.2 μm nominal | 0.2 μm absolute | 0.1 μm absolute |
| (polypropylene or glass) | (glass or cellulose) | (glass or cellulose) | integrity testable | integrity testable |

**FIGURE 19**   Filter train for producing sterile human or mammalian serum.

serum filtrations, a 0.1-$\mu$m sterilizing grade filter is sometimes used as the final filter to increase the removal of any mycoplasma contamination that may be present in the serum.

Generally a great deal of thought and testing is given to the selection of the final (sterilizing grade) filter in the train, however if you are required to produce a sterile effluent, your choices on the type ($\mu$m-size) of filter cartridge is limited to those membrane filters and filter cartridges that meet the HIMA and FDA regulatory guidelines that define a "sterilizing grade filter." There are, however, different polymeric materials used for the various sterilizing grade filters that are available and those initially selected should be evaluated with your process solution for filter compatibility and/or product degradation or adsorption. Once you have chosen an appropriate final filter cartridge, the next task is to choose the most appropriate clarifying and/or prefilters to extend the lifetime (capacity) of the final filter cartridge.

## 2. Filter Sizing

Choosing the appropriate prefilter combinations can be the key to filter train optimization and cutting overall filtration costs. However, with the number and type of prefilter cartridges available, choosing the best prefilter train offers a unique challenge. In order to meet this challenge, two procedures, (a) flow decay trials, and (b) $V_{max}$ trials, are generally employed.

*a. Flow Decay.* Flow decay is the traditional method used to measure the flow rate and volume that will pass through a filter over a prescribed time period, generally until flow drops to 20% or 10% of the initial flow rate. Flow decay trials are run on small area filters, then scaled-up to determine filter cartridge capacity. The effluent from the small area filter is then used to run a flow decay test on the next prefilter or the final filter in the filter train. By evaluating various prefilter media one can determine the best filter combination for protecting the final filter. The diagram in Figure 20 shows the apparatus required to run the flow decay test.

*1. Flow Decay Procedure*:   In the flow decay procedure filter disks (generally 47mm) are challenged with a representative sample of the process solution that will be used in the large scale production runs. Tests are run at a constant upstream pressure until the flow rate to the disk declines to 20 or 10% of the initial flow rate. Final sterilizing grade filters of 0.2 and 0.1 $\mu$m and membrane based prefilter material are run at 20 to 30 psid or at the differential pressure where the cartridge will be changed in the filter application. The more open clarifying filters and depth prefilters are run at 5 to 10 psid. The filter with the highest retention in the train is run first, in

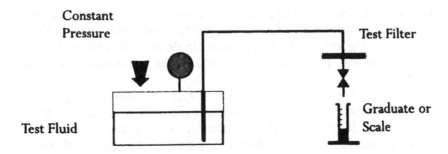

FIGURE 20 Apparatus used to perform flow decay or $V_{max}$ trials.

order to assess the effectiveness of various prefilter materials that will be tested to extend the life of this final filter. The total volume fed to each disk is recorded. To scale up the flow decay data on the disk to a cartridge, use the equation:

$$\frac{\text{Cartridge Filter Area}}{\text{Disk Filter Area}} \times (\text{Volume Fed to Disk}) =$$

$$\text{Volume cartridge will process}$$

*b. $V_{max}$* $V_{max}$, a modified flow decay test based on the gradual pore blocking mechanism is used to predict the maximum volume that will pass through a filter at infinite time (Badmington, 1993; Brose et al., 1994; Badmington et al., 1995). The flow rate and cumulative volume through a filter disk is recorded for 10 minutes, at one minute increments. A straight line plot of time (minutes) / volume (liters) versus time (minutes), modeled from the gradual pore blocking mechanism, can be used to predict the total volume through a filter at an 80, 90 or 100% decline in the initial flow rate. Usually at an 80 to 90% reduction in initial flow rate a filter cartridge would be considered beyond its useful life. To calculate the volume of solution that will pass through the test filter as flow rate is reduced, use the equations listed below:

$V_{max}$ at 100% flow loss ($V_{max}$) $\quad = \quad$ 1/ slope of the straight line plot

$V_{max}$ at 80% flow loss ($V_{80\%}$) $\quad = \quad$ 0.55 × $V_{max}$

$V_{max}$ at 90% flow loss ($V_{90\%}$) $\quad = \quad$ 0.68 × $V_{max}$

As discussed in section two of this chapter on basic filtration theory, most biological solutions contain small deformable particles that will typically block filters by gradual pore blocking and therefore, allow $V_{max}$ data to predict the capacity of a filter. $V_{max}$ testing is extremely useful in sizing and optimizing filtration trains in that it can dramatically reduce the time required to test the capacity of a disk over the traditional flow-decay method.

*1. $V_{max}$ Procedure*:    To perform a $V_{max}$ test, filter disks (generally 47mm in diameter) are challenged with a representative sample of the process solution. Tests are run at a constant upstream pressure for 10 to 15 minutes, recording the cumulative volume through the filter at 1.0 minute increments. The same test apparatus shown in Figure 20 for developing flow decay can be used for $V_{max}$ trials. 0.2 $\mu$m final filters and membrane based prefilters are generally run at 30 psid. Depth prefilters are run at 5 to 10 psid. The filter with the highest retention in the train is run first, in order to assess the effectiveness of various prefilter materials that will be tested to extend the life of this final filter. The total volume fed to each disk is recorded. To determine the flow decay for a cartridge use equation:

$$\frac{\text{Cartridge Filter Area}}{\text{Disk Filter Area}} \times (\text{Volume Fed to Disk}) =$$

$$\text{Volume cartridge with process}$$

An example of how the flow decay or $V_{max}$ sizing approach can optimize a filtration train is summarized in Table 6. Based on the test data developed on 47-mm disks, and then scaled up to 10-in. cartridge equivalents, filtration train #3, containing two prefilters before the final filter, would offer a significant reduction in the filtration cost to process 1000 l of 5% calf serum, compared to filter train #1 or #2.

As an additional benefit to filtration train optimization, $V_{max}$ testing can be used by Production and Quality Control managers to monitor the batch to batch consistency of their process solution. In some instances, where the quality of process solutions may vary widely, a 10-minute $V_{max}$ test can easily predict if premature filter plugging is expected in this process filtration step.

To be effective in sizing filtration systems it is important to:

1. Understand your overall objectives.
2. Determine what is critical in your separation and answer the questions listed in Figure 21 for your process solution characteristics and filtration conditions.
3. Record accurate data with your test solutions and be sure that your test solution is representative of your final process solution.

**TABLE 6**   Filtration Train Optimization to Process 1000 Liters of 5% Calf Serum

| | 1.2 $\mu$m Nominal cellulose prefilter | 0.2 $\mu$m Nominal cellulose prefilter | 0.2 $\mu$m PVDF final filter | Total relative cost |
|---|---|---|---|---|
| **Train 1 with final filter only** | | | | |
| Capacity on a 47mm disk (liters) at 80% reduction in initial flow rate or $V_{80\%}$ | | | 0.26 | |
| Disk capacity scaled to a 10" cartridge (liters) | | | 130 | |
| Number of 10" cartridges required to process 1000 liters | | | 8 | 100% |
| **Train 2 with prefilter cartridge, then final filter** | | | | |
| Capacity on a 47mm disk (liters) at 80% reduction in initial flow rate or $V_{80\%}$ | | 0.46 | 3.3 | |
| Disk data scaled to a 10" cartridge (liters) | | 230 | 1600 | |
| Number of 10" cartridges required to process 1000 liters | | 5 | 1 | 44% |
| **Train 3 with 2 prefilter cartridges in series then final filter** | | | | |
| Capacity on a 47mm disk (liters) at 80% reduction in initial flow rate or $V_{80\%}$ | 8.0 | 6.8 | 4.2 | |
| Disk data scaled to a 10" cartridge (liters) | 4000 | 3400 | 2100 | |
| Number of 10" cartridges required to process 1000 liters | 1 | 1 | 1 | 25% |

Reprinted from Badmington, 1993 with permission.

4. Design a preliminary cartridge filtration train based on your sizing data, see Figure 22.
5. When sizing large scale filtrations (those requiring cartridges greater than 10 ins. in length), perform a pilot scale run with 2- or 4-inch cartridges before running the full scale trial.

PROCESS SOLUTION:

Composition: _____

Viscosity: _____

Surface Tension: _____

Particle Distribution _____

FILTRATION CONDITIONS:

Constant Pressure or Constant Flow Rate   (circle one)

Sterilization of Filters:  Steam in Place (SIP), Autoclave or Not Required   (circle one)

Batch Size: _____

Flow rate: _____

Temperature: _____

Initial Differential Pressure Requirements:_____

Final Differential Pressure Requirements: _____

Required Filtration Time: _____

FIGURE 21    Initial information needed to set up a well designed filtration train.

| Filter Train Cartridge # | Type of Filter (clarifying, prefilter) | Micron Rating | Length (inches) | Number Required |
|---|---|---|---|---|
| #1 | | | | |
| #2 | | | | |
| #3 | | | | |
| Final Filter (Sterilizing) | | | | |

FIGURE 22    Preliminary filter cartridge information for a specific application based on disk sizing data.

6. Rely on technical assistance and testing help from the filter manufacturers.

## B. Gases

The Aseptic Guideline (US Dept. of Health and Human Services, 1987) requires the use of sterilizing grade filters for critical contact gases. This would include the filter cartridges used on water for injection (WFI) tank vents, sterile product tank vents, bio-reactor feed gas lines, lyophilizers, autoclaves, and biotechnology fermentation tank vents. Sterilizing grade gas filter cartridges must be integrity testable, sterilizable, nonfiber releasing, and be bacterial retentive in dry and wet air. Ideally, they will have a low pressure drop, both initially and throughout their useful cycle, the ability to be repeatedly steam sterilized, the ability to withstand long term high temperature air exposure, and be viral retentive in dry and moist air. Due to the compressibility of gases at various pressures, the sizing of air filters for high flow applications can be complex. Filter manufacturers generally have filter sizing programs available to assist you with this work. To evaluate the overall economics in the sizing of a gas filtration system, the user should consider the labor for sterilizing, integrity testing, and changing the filters, as well as the initial cost of the filters and housings (Keating et al., 1992).

## VIII. FILTER APPLICATIONS SPECIFIC TO PROTEIN PROCESSING

### A. Concentration

To concentrate protein solutions, Tangential Flow Filtration techniques are often employed. Concentration is typically carried out with ultrafiltration membranes of a nominal molecular weight limit (NMWL) of 10,000 to 100,000 Daltons. Membranes with greater than 99% retention of the target protein are necessary for high protein yield. Modules with turbulence-promoted channels are preferred because of their high flux and low circulation rates with protein solutions. Figure 23 shows a typical tangential flow filtration process for the concentration of a protein solution.

### B. Diafiltration (Washing)

In diafiltration processes, TFF is used to change the composition of the supporting protein buffer, without changing the protein concentration. As illustrated in Figure 23 diafiltration solution is added to the feed tank at the same rate the permeate is removed. The volume of removed permeate equals the volume of added diafiltration solution. Assuming zero retention

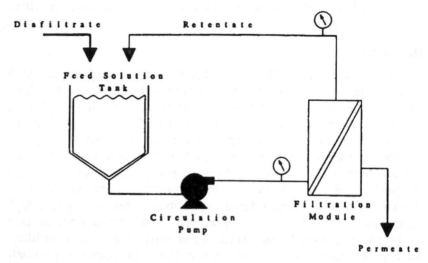

Example:

|                        |                        |
|------------------------|------------------------|
| Membrane area          | 50 m$^2$               |
| Average Permeation Rate | 2,500 liters per hour |
| Circulation Rate       | 25,000 liters per hour |

Process Description

| Time (hours) | Volume in Feed Tank (liters) | Conc. of Protein in Feed Tank | Volume permeated (liters) |
|--------------|------------------------------|-------------------------------|---------------------------|
| 0            | 11,000                       | 1.0 grams/liter               | 0                         |
| 4            | 1,000                        | 10.8 grams/liter              | 10,000                    |

Result: Protein in feed at end of concentration is 98% of original protein

FIGURE 23   Tangential flow filtration processes for concentrating protein (reprinted with permission from Millipore).

of the buffer solutes, a diafiltration volume equal to 5 times the volume of the original solution is required to remove 99% of the initial buffer solute. Diafiltration is commonly used to remove precipitation solvents and to adjust the buffer composition before or after the chromatographic steps in a protein refolding process.

## C. Clarification

In the protein clarification process, the protein solution is passed through the filter membrane, while cells, larger molecules, and cell debris are retained. During the separation of protein products from large mammalian cells like Chinese Hamster Ovary (CHO) cells or large particulate cell debris, both cartridge depth filters with nominal ratings in the 5 micron range and TFF systems with microporous membranes can be employed. Both systems bring advantages and disadvantages to this type of separation. After the large cells are removed by either a depth filter or a TFF system, a prefilter is generally required to extend the life of the sterilizing-grade final filter cartridge or to protect a columnn that may be used in the separation process. The selection of the best system can only be made after reviewing (1) the objective of the separation, (2) the characteristics of the process fluid, and (3) the overall processing costs.

## D. Pyrogen Removal

Standard 0.2- and 0.1-$\mu$m rated sterilizing grade filters will not remove pyrogenic lipopolysaccharides from protein solutions. However, charged modified 0.2- and 0.1-$\mu$m filter cartridges run in the deadend filtration mode have been used for these applications. The charged filter removes the pyrogenic material by adsorption at the charged sites. For pyrogen removal to be based on size exclusion, ultrafiltration membranes of 10,000 to 100,000 Daltons are required. Reverse Osmosis (RO) membranes are also used for pyrogen removal.

## ACKNOWLEDGEMENTS

The author would like to acknowledge the assistance given by Millipore in providing the product photographs and the time that was required to prepare this chapter. Thanks are also due to Robin De La Parra for providing the scanning electron micrographs and Dave Reader for assistance with the system flow diagrams.

## REFERENCES

American Society for Testing and Materials. Standard Test Method for Determining Bacterial Retention of Membrane Filters Utilized for Liquid Filtration. Annual Book of ASTM Standards. West Conshohocken, PA: 1988, F838-83.
Australian Dept. of Community Services and Health, Therapeutic Goods Administration. Code of Good Manufacturing Practice for Therapeutic Goods—Medicinal Products. January, 1992. 136 Narrapundah Lane, Symonston, ACT.
Baccaro M. Membrane Processes. Pharmaceutical Engineering 13:8–16, 1993.
Badmington F. Reducing Filtration Costs Through Prefilter Optimization. Poster Session. Parenteral Drug Association Annual Meeting, Orlando, FL, 1993.

Badmington F, Honig E, Payne M, Wilkins R. Vmax Testing for Practical Microfiltration Train Scaleup in Biopharmaceutical Processing. Pharm Technol 19: 64–76, 1995..

Brose DJ, Cates S, Hutchison FA. Studies of the Scale-Up of Microfiltration Membrane Devices. J Parenteral Sci Technol 48:184–188, 1994.

Brose DJ, Henricksen G. A Quantitative Analysis of Preservative Adsorption on Microfiltration Membranes. Pharm Technol 18:65–72, 1994.

Bowen RW, Gan Q. Properties of Microfiltration Membranes: The Effects of Adsorption and Shear on the Recovery of an Enzyme. Biotechnology and Bioengineering 40:491–497, 1992.

Commission of the European Communities. The Rules Governing Medicinal Products in the European Communities, Vol. IV. Good Manufacturing Practice for Medicinal Products, 1992. 3 Dag Hammarskjold Plaza, New York, NY.

Datar R, Martin JM, Manteuffel RL. Dynamics of Protein Recovery from Process Filtration Systems Using Microporous Membrane Filter Cartridges. J Parenteral Sci Technol 46:35–42, 1992.

Dickerson C. Filters and Filtration Handbook. 3rd ed. Elsevier, 1992. 52 Vanderbilt Ave, New York, NY.

Emory S. Principles of Integrity Testing Hydrophilic Microporous Membranes. Part I. Pharm Technol 13:68–77, 1989(a).

Emory S. Principles of Integrity Testing Hydrophilic Microporous Membranes. Part II. Pharm Technol 13:36–46, 1989(b).

Grace HP. Structure and Performance of Filter Media. A I Ch E Journal 3:307–336, 1956.

Health Industry Manufacturers Association (HIMA). (1982) Microbiological Evaluation of Filters for Sterilizing Liquids. HIMA Doc no 3, vol 4, 1982.

Hermans PH, Bredee HL. Rec Trav Chim des Pays-Bas. 54:680, 1935.

Hermans PH, Bredee HL. Principles of the Mathematical Treatment of Constant Pressure Filtration. J Soc Chem Ind, Trans Commun 55:1–4, 1936.

Hermia J. Constant Pressure Blocking Filtration Laws, Application for Power-Law Non-Newtonian Fluids. Trans I Chem E 60:183–187, 1982.

Hu H, Camilleri J, Tamashiro W. A New Membrane for Biopharmaceutical Filtration. Pharm Technol 17:32–44, 1993.

Johnston PR. Fundamentals of Fluid Filtration...A Technical Primer. Littleton, CO: Tall Oaks, 1990.

Keating P, Levy R, Payne M, Proulx S, Rowe P, Pearl S. Performance Testing of Membrane-Based Filters Used in the Filtration of Industrial Fermentation Air, Biopharm 5:36–41, 1992.

Leahy TJ. Validation of Bacterial Retention by Membrane Filtration: A Proposed Approach for Determining Sterility Assurance. PhD Dissertation, University of Massachusetts, Amherst, MA, 1983.

Levy RV. The effect of pH, Viscosity and Additives on the Bacterial Retention of Membrane Filters Challenged with Pseudomonas Diminuta. Fluid Filtration:

Liquid Vol II, American Society for Testing Materials, Special Technical Publication, 1987, 975: pp. 80–89. West Conshohocken, PA.

Martin JM, Manteuffel RL. Protein Recovery from effluents of Microporous Membrane Filters. Biopharm 1:20–27, 1988.

Meltzer TH. Filtration in the Pharmaceutical Industry. New York: Marcel Dekker, 1987.

Meltzer TH. Filtration: A Critical Review of Filter Integrity Testing: Part I. the Bubble Point Method, Assessing Filter Compatibility, Initial and Final Testing. Ultrapure Water 6:40–51, 1989.

Meltzer TH. Filtration: A Critical Review of Filter Integrity Testing: Part II. the Diffusive Air Flow and Pressure-Hold Methods, Assessing Filter Compatibility, Initial and Final Testing. Ultrapure Water 6:44–56, 1989.

Meltzer TH. The Insufficiency of Single Point Diffusive Air Flow Integrity Testing. J Parenteral Sci Technol 46:19–21, 1992.

Millipore Annual Report. Millipore Corp, Bedford, MA, 1996.

Nema S, Avis K. Loss of LDH Activity During Membrane Filtration. J Parenteral Sci Technol 47:16–21, 1993.

Olson WP. Separations Technology Pharmaceutical and Biotechnology Applications. Prairie View, IL: Interpharm, 1995.

Pall Annual Report. Pall Corp, East Hills, NY, 1996.

Pitt A. The nonspecific Binding of Polymeric Microporous Membranes. J Parenteral Sci Technol 41:110–113, 1987.

Rousseau RW. Handbook of Separation Process Technology. New York: John Wiley, 1987.

Rubow KL. Submicron Aerosol Filtration Characteristics of Membrane Filters. Ph.D. Thesis, University of Minnesota, Minneapolis, MN, 1981.

Rubow KL, Liu BYH. Characterization of Membrane Filters for Particle Collection Fluid Filtration: Gas. In: Raber RR, ed. Volume 1 ASTM STP 975. Philadelphia: American Society for Testing and Materials, 1986.

Rubow KL, Liu BYH, Grant D. Characteristics of Ultra-High Efficiency Membrane Filters. Proceedings of the 33rd Annual Meeting of the Institute of Environmental Sciences, 1987, pp 383–387.

Sarry C. Sucker H. Adsorption of Proteins on Microporous Membrane Filters. Part I. Pharm Technol 16:16–23, 1992(a).

Sarry C, Sucker H. Adsorption of Proteins on Microporous Membrane Filters. Part II. Pharm Technol 16:38–48, 1992(b).

Segers P, Vancanneyt M, Pot B, Torck U, Hoste B, Dewettinck D, Falsen E, Kersters K, De Vos P. Classification of *Pseudomonas diminuta* Leifson and Hugh 1954 and *Pseudomonas vesicularis* Busing, Doll and Freytag 1953 in *Brevundimonas* gen. nov. as *Brevundimonas diminuta* comb. nov. and *Brevundimonas vesicularis* comb. nov. Respectively, International J. Systematic Bacteriol 44: 499–510, 1994.

Stone T, Goel V, Leszczak J. Methodology for Analysis of Filter Extractables: A Model Stream Approach. Pharm Technol 18:116–130, 1994.

Truskey GA, Gabler R, DeLeo A, Manter T. The Effect of Membrane Filtration Upon Protein Conformation. J Parenteral Sci Technol 41:180–193, 1989.

United States Pharmacopeia, USP 23, National Formulary, NF18. United States Pharmacopeial Convention, Rockville, MD, 1995.

US Dept of Health and Human Services, Public Health Service, Food and Drug Administration. Guideline on Sterile Drug Products Produced by Aseptic Processing, Rockville, MD, 1987.

Van den Oetelaar JM, Mentink IM, Brinks JB. Loss of Peptides and Proteins upon sterile Filtration Due to Adsorption to Membrane Filters. Drug Development and Industrial Pharmacy 15:97–101, 1989.

Weismantel GE. Filtration and Separation. Chem Engineering 57–87, 1986.

Winston HO WS, Sirkar KK. Membrane Handbook. New York: Chapman & Hall, 1992.

# 8

## Considerations for Elastomeric Closures for Parenteral Biopharmaceutical Drugs

JOHN A. BONTEMPO

Biopharmaceutical Product Development, East Brunswick, New Jersey

## I. INTRODUCTION

This chapter is essentially an overview on the interactions of proteins and peptides with elastomeric closures rather than with glass containers. However, what is highly recommended, from personal experience, is that with both liquid or lyophilized sterile dosage forms, Type I glass is the only type that should be used. This type, composed principally of silicon dioxide and boric oxides, as classified by the United States Pharmacopeia (USP) possesses the highest chemical resistance due to low reactivity, low leachability, and low thermal coefficient of expansion.

Selection of more than one elastomeric closure is of critical importance by the formulation scientist. The key parameters that must be considered will be addressed in this chapter since the selections of the elastomeric closures may have a significant impact on the shelf life of the future marketable products.

These are some of the desirable properties that elastomeric closures should have for one's own specific product, such as

- Essentially nonreactive physically and chemically with the complete formulation ingredients
- Complete barrier to vapor/gas permeation
- Penetrable with a hypodermic syringe needle, or a plastic spike for IV administration
- Compressibility and resealability
- Resistant to coring and fragmentation
- Maintains seal interface and package integrity

Unfortunately all these properties are not found in any one type of elastomeric closure today; consequently, it is very important to select an elastomer with as many of the desired properties as one can (1).

### A. Application of Elastomeric Closures for Some of the Most Often Used Primary Packaging Systems

The selection of the primary packaging will also require a specific type of closure. There are numerous packaging systems for parenteral products and the following are some of the most often used.

- Vial closures for small volume parenterals (SVP)
- Vial closures for large volume parenterals (LVP)
- Intravenous set components
- Intravenous solution closures
- Closures for liquid, lyophilized, and powder-filled products
- Sterile empty syringes and prefilled syringes
- Multicomponent syringes and multicomponent vials

## II. PHYSICAL PROPERTIES OF RUBBER

A pharmaceutical formulator should be quite knowledgeable in selecting the elastomeric closures. A number of physical properties have been identified as being important in making the appropriate selection (2,3). Elastomer, or so called rubber, closure is composed of polymers that, at room temperature, can be stretched considerably and, upon release, returns to its original shape (4). Rubber is a very complex material made of up to ten or more raw materials and the chief polymeric component of rubber is the elastomer (5). The words "elastomer" and "rubber" are used interchangeably. The following characteristics are most appropriate to consider in selecting the rubber closure.

1. *Compression.* This is a measure of resiliency of the rubber, an important property for resealability.
2. *Coring.* This is a measure to resist fragmentation of the rubber during puncture. This property must be well-taken into consideration when a multidose is selected by the pharmaceutical company.
3. *Moisture vapor transmission (MVT).* It is important to have the proper elastomer to form a complete barrier to prevent entrance of any vapors or reactive gases into the container, especially when working with lyophilized products or hygroscopic substances.
4. *Insertion force.* The rubber closure should not be so rigid as to require excessive pressure to insert a hypodermic needle or a plastic spike for IV administration sets.
5. *Durometer hardness.* The measure of resistance to indentation.

6. *Resealability.* The rubber closure should be such as to allow elasticity upon insertion and withdrawal of a hypodermic needle and sufficient elasticity to provide a close fit between the closure and the lip and neck of the vial.
7. *Vacuum retention.* The elastomeric closure should maintain the selected level of vacuum and it could be a stability factor in terms of a closure/container system (6).
8. *Solvent resistance.* The ability of the elastomer to withstand decomposition or disintegration when immersed in solvents or medical solutions for a specific time and temperature.

## A. Classification of Elastomers

Elastomeric closures are made from two types of elastomers or rubbers. One is the *thermoset rubber* which undergoes a chemical reaction during the molding step. When this happens there is an insertion of crosslinks or bonds between the long polymer chains to form a resilient three dimensional network. Once an elastomeric closure is molded, it cannot be changed into another shape (7).

The other type of rubber closure is called *thermoplastic rubber* (8). These closures are essentially similar to conventional plastics, such as polyethylene or polystyrene. The process of making these closures is reversible. There are no chemical reactions involved in these closures.

The crosslinking process is not a chemical reaction, it is a physical intertwining of polymeric chains. This three dimensional network gives thermoplastic closures their elasticity and resiliency. Elastomers are classified as unsaturated, based on the number of reactive double bonds in the main chain or side chain on the elastomer. The degree of unsaturation determines both the physical and chemical properties of the elastomer which, in turn, greatly influences the properties of the rubber formulation. Table 1 lists the elastomers which may be used in pharmaceutical closures (9).

## B. Composition of Rubber Closures

In addition to the basic polymers, natural rubber, and synthetic polymer, elastomeric closures are formulated with many ingredients. These ingredients are selected according to their functions.

1. *Vulcanizing agents* such as sulfur, phenolic resins, and organic peroxides. These are chemicals used to crosslink elastomeric chains

TABLE 1 Saturated and Unsaturated Elastomers

| Saturated | Unsaturated |
| --- | --- |
| Butyl (IIR) | Natural (NR) |
| Halogenated butyls, chloro and bromo (CIIR) and (BIIR) | Styrene butadiene (SBR) |
| Ethylene–propylene rubber (EPM) | Polysoprene (IR) |
| Ethylene–propylene–diene rubber (EPDM) | Nitrile (NBR) |
| Silicone (Q) | Neoprene (CR) |
| Urethane (U) | Polybutadiene (BR) |
| Fluoroelastomers (FKM) | |

Physical, chemical properties and chemical structures are found in Chapter 11, Elastomeric Closures for Parenterals, by Smith, Nash (Ref. 5).

into the three dimensional network needed to give the rubber formulation the desired physical and chemical properties.

2. *Accelerators* such as amines, thiols, sulfamides, ureas, and thiuram compounds. These reduce the cure time by increasing the vulcanization rate.
3. *Activators* such as zinc oxide, zinc stearate, and stearic acid. These increase the rate of crosslinking.
4. *Antioxidants* such as amines, diethiocarbamates, and paraffin waxes.
5. *Fillers* such as aluminum and calcium silicates, carbon black, titanium dioxide, and barium sulfate.
6. *Lubricating agents* such as paraffin, mineral oils, fatty oils, and organic phosphates.
7. *Pigments* such as carbon black, chromium oxide, and iron oxide. These are used for aesthetic or functional purposes.

Other acceptable materials can be found in the Code of the Federal Regulations Sec. 177.2600 or in Technical Methods Bulletin No. 1 of the PDA (7).

The elastomer and additives are combined into homogeneous mass which is then vulcanized into a desired closure shape.

Today there is no single rubber closure on the market that can be described as a perfect formulation. *There exists needs to improve polymers, especially natural rubber.* There is no single rubber formulation that meets all the requirements of the pharmaceutical packaging industry (10).

The pharmaceutical rubber closures companies are keenly aware for the need of a much improved product closure, especially for the biopharmaceuticals being developed today. Significant developments along these lines are being addressed by these companies.

## III. POTENTIAL INTERACTIONS OF DRUG FORMULATIONS WITH RUBBER CLOSURES

As stated previously, there is no rubber closure formulation which can be considered inert. There are always some levels of reactivity, however small. The selection of the rubber closure formulation is influenced by several factors such as (1) buffer system, (2) color preference, (3) configuration of closure, (4) drug active substance, (5) metallic sensitivities, (6) method of sterilization, (7) moisture vapor/gas protection, (8) pH of the formulation, (9) preservatives, (10) solvent vehicle, (11) doses per container, and (12) configuration of closure. Consequently, a pharmaceutical formulator should evaluate the existing rubber formulations and select two or more that would protect and preserve the drug formulation through the various stress conditions of stability studies. These stress conditions may induce physicochemical interactions which might render the formulation adulterated.

Scheme No. 1 illustrates how ingredients of a formulation can undergo interactions between the active drug substance and the primary packaging system (11).

Rubber closures will react with the drug formulation in several ways, especially under the stress of laboratory stability tests simulating environmental shipping and receiving conditions.

## A. Adsorption and Absorption

*Adsorption* is the phenomenon of a drug substance concentrating on the surface of the rubber closure.

*Absorption* is the process of a drug substance penetrating through the rubber closure surface and migrating throughout its mass.

Both adsorption and absorption can occur in a drug formulation. Some of the factors influencing *sorption* (adsorption and absorption) are (1) effect of concentration, (2) partition coefficient, (3) pH of the solution, (4) effect of excipients, (5) effect of temperature, (6) structure of polymeric sorbent, and (7) structure of the sorbate (11).

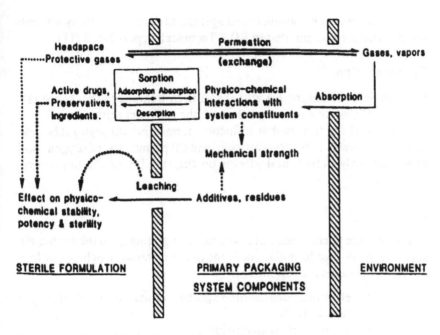

SCHEME 1　Potential physicochemical interactions.

## B. Sorption of Antimicrobial Agents by Rubber Closures

One of the paramount concerns of the pharmaceutical formulator in selecting an antimicrobial substance for a multidose formulation is the potential interaction of the antimicrobial agent with the rubber closure.

The decrease of preservative(s) effectiveness by sorption can have a significant negative effect on shelf life of the product (12–14). The preservative is the major consideration in the formulation.

From personal experience, the loss of benzyl alcohol and phenol occurs over a period of time in controlled stability studies and the concentrations decrease to a level that is no longer effective enough to retain their Antimicrobial Effectiveness Test compliance with the United States Pharmacopeia. Some of their loss was by absorption into the rubber closure as well as some by volatilization at various temperature levels.

If the liquid formulation in a sealed vial is stored inverted, there occurs a faster and greater absorption, which is also related to temperature.

An excellent review of antimicrobial agents and their interactions with rubber closures are documented in PDA Technical Report No. 5 (11).

## C. Permeation

*Permeation* is the adsorptive and absorptive phenomena that eventually leads to diffusion of a substance through the rubber closure.

Some of the factors that influence permeation through rubber closures are (1) water vapor transmission, and (2) permeation of oxygen. Both have extremely detrimental effects on the stability of the active drug substance.

## D. Leaching

*Leaching* is the phenomenon of a substance migrating from the rubber closure into the dosage formulation. Some of the materials leaching out from the rubber closures are residues of:

(1) Accelerators, such as mercaptobenzothiazole and tetramethylthiuram disulfide
(2) Activators, such as zinc oxide
(3) Lubricants such as stearic acid as well as other inert components
(4) Antioxidants, such as hindered phenols

Detailed procedures for the extraction, isolation, and identification of extractables from rubber closures are found in the "Extractables from Elastomeric Closures: Analytical Procedures for Function Group, Characterization and Identification," published in the PDA Methods Bulletin No. 1.

The absorption of preservatives such as benzyl alcohol, methylparaben, and chlorobutanol have been studied. Natural rubber closures had the least amount of extractables (15). Other extractables such as fillers, have also been reported that have a negative impact on the formulation (16–19).

## IV. CONTROL AND TESTING OF ELASTOMERIC CLOSURES

The various classes of raw materials in the manufacturing of pharmaceutical rubber formulations were discussed previously in this chapter; however, there are over 10,000 materials listed in the rubber Blue Book, published annually (20). Although it is generally accepted that rubber is somewhat inert, on the contrary it is found to be reactive and the number of ingredients that are generally recognized as safe (GRAS) is quite small.

The regulations governing these ingredients are the regulations that apply to foods (21). The following are applicable sections of the Code of the Federal Register (22).

21 CFR 175.  Indirect food additives, adhesive coatings, and components
21 CFR 177.  Indirect food additives and polymers
21 CFR 178.  Indirect food additives, adjuvants, production aids, and sanitizers
21 CFR 182.  Substances generally recognized as safe (GRAS)
21 CFR 184.  GRAS direct food additives
21 CFR 185.  GRAS indirect food additives

## A. Identity Testing

In order to establish acceptance of rubber closure lots, a pharmaceutical company sets up specifications in order to assure uniformity from lot to lot to lot. Some of the recommended tests are:

1. Specific gravity or density to determine the filler content.
2. Determination of ash, to also ascertain the filler content of the rubber closure. Specific gravity and percentage ash values are used by formulation scientists to identify vulcanized rubber formulations (23–26)
3. Infrared spectrometry to yield identity of the polymer.
4. Ultraviolet spectroscopy to identify extractables, curing agents, accelerators, and antioxidants.
5. Visual and olfactory tests which yield specific color and smell, depending on the composition of the elastomer formulation.

Other physical and chemical tests are turbidity, reducing agents, heavy metals, pH change, and total extractables (24,27,28,29). These are the required USP tests and must be in the DMF of the rubber manufacturer.

## B. Biological Testing of Elastomers

Rubber closures are required to be tested in vitro and in vivo, according to the USP XXIII, Biological Tests Protocol, Sections 381, 387 and 388.

The in vitro tests focus on the reactivity of mammalian cell cultures to rubber and rubber extracts via direct contact, elution, or agar diffusion. The in vivo tests focus on the biological responses of animals to rubber extracts via systemic injection and intracutaneous reactivity tests. Other tests include pyrogen testing, cytotoxicity tests, hemolysis, and bacterio-static and bactericidal tests.

## C. Washing of Rubber Components

Washing rubber closures for parenteral application requires several con-siderations in order to minimize surface particles, endotoxins, and other possible contaminations (30). In selecting a washer, the design should be one of the following: (1) overflow with air agitation, (2) revolving drum, (3) flotation, and (4) rotating cages. Detergent is often used and the choice is made case by case. This is based on previous experience. Detergents may be tetrasodium pyrophosphate or trisodium phosphate. The washing should be a gentle agitation so that no additional particulate matter is gen-erated. There should be a detailed and very specific washing procedure for each rubber closure. Rubber closures should be examined following wash-ing for physical appearance, shape aberration, particulate matter, residue, and for changes in physicochemical tests. Stoppers should be sterilized, dried, and properly packaged for their intended use. After drying, physical appearance, dimensional characteristics, and residual moisture must be ad-dressed. Methods of sterilization that can be used for rubber closures are steam, radiation, and ethylene oxide. Each method of sterilization should be considered for its effect on the physical appearance, dimensional char-acteristic, mechanical, and chemical interaction.

Moisture residues are a factor that must be considered in a finished formulation. There are essentially three factors governing the residual moisture and these should be evaluated very carefully. First is the nature of the rubber closure itself; secondly, the sterilization cycle and thirdly, the drying cycle. The residual moisture is of a particular interest in terms of its effect on the biological activity and long term stability of the product. Lyo-philized products of biopharmaceuticals must be especially monitored and their moisture to stability relationships determined. The level of residual moisture for optimal retained biological activity of a product must be ascer-tained case by case. Residual moisture determination methods most often applied for this purpose are the gravimetric and the Karl Fischer. Of the two, the Karl Fischer titration method is generally the most accurate.

## V. INTERACTION OF PROTEINS AND PEPTIDES WITH ELASTOMERIC CLOSURES

As stated earlier this chapter, it is important that the selection of the rubber stopper materials for each formulation should be made only after a prescreening has been performed on the effect(s) of these materials on the formulation. When the active drug substances are proteins and peptides, the major considerations of a rubber closure should focus on the adsorption of these active drug substances on the surface of the closures. The extended shelf life of these products will depend on how well the product had been protected during storage under varying environmental conditions. Each protein is a case by case study. The environmental conditions selected for one protein may or may not be suitable for another. The primary goal again, for a formulation scientist, is to assure (1) the biological efficacy, and (2) the safety of the product.

## A. Prescreening Testing of Rubber Closures

The requirements for glass have been addressed previously. At this point, interest focuses on the rubber closures. The pharmaceutical industry, for the most part, has been using butyl, halobutyl, natural, and polyisoprene elastomers for their products (31). The elastomeric closure companies are focusing on coating some of their rubber closures with a very thin film of various inert polymers in order to achieve greater compatibility, flexibility, low levels of particulates, and machinability (32–34).

A prescreening procedure (31) would allow formulation scientists to select the most suitable rubber stopper formulation from the data of the prescreening studies. These should be some of the pertinent considerations.

- Acquire all technical information from your rubber stopper vendors on more than one potential rubber formulation.
- Acquire, as much as available, physicochemical properties of active drug substance.
- Select a minimum of three closures, more if possible. Some of these should be coated with an inert polytetrafluoroethylene film or with a fluorinated coating.
- Design, based upon the physicochemical properties, an accelerated realistic stability study for one to two months. Include in this study effects of temperature, light, pH, position of vials in an upright and inverted position, closure, and formulation changes, like closure extractables, color, precipitation, and pH.

Prior to initiating this study, it is of paramount importance to have initially developed an analytical quantitative assay, measuring plausible quantitative activity, and later follow this with another method. One method of assay is not sufficient to be used as a stability indicator assay. Once reproducible results of these studies are obtained, the formulation

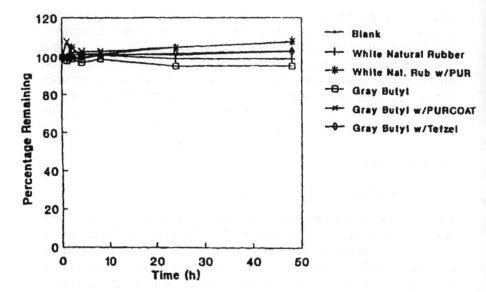

**FIGURE 1** Protein adsorption: various stoppers with IgG. (From Kiang, Wang, Kao. PDA Short Course, Oct, 1995, Malvern, PA.)

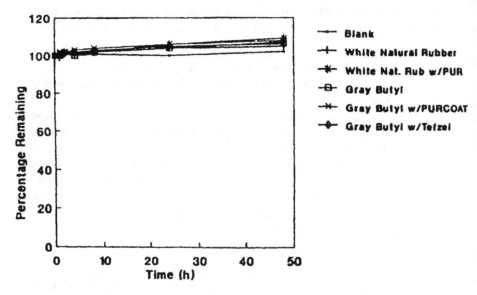

**FIGURE 2** Protein adsorption: various stoppers with gastrin. (From Kiang, Wang, Kao. PDA Short Course, Oct, 1995, Malvern, PA.)

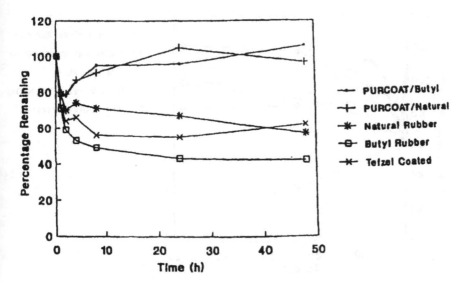

FIGURE 3   Protein adsorption: various stoppers with PTH. (From Kiang, Wang, Kao. PDA Short Course, Oct, 1995, Malvern, PA.)

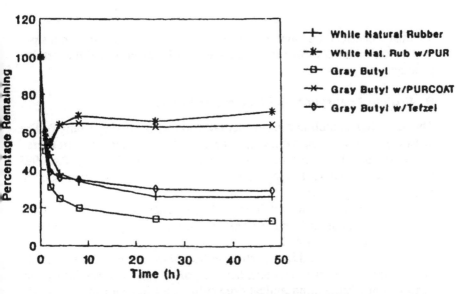

FIGURE 4   Protein adsorption: various stoppers with ACTH. (From Kiang, Wang, Kao. PDA Short Course, Oct, 1995, Malvern, PA.)

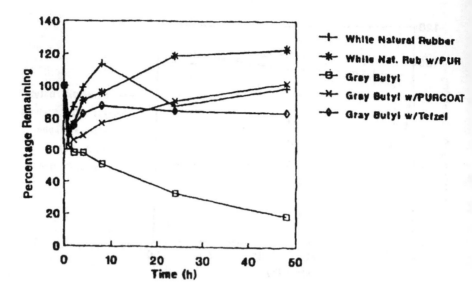

**FIGURE 5** Protein adsorption: various stoppers with pins. (From Kiang, Wang, Kao. PDA Short Course, Oct, 1995, Malvern, PA.)

scientist will select a minimum of three stoppers to incorporate in his potential formulation(s) for clinical studies, as well as potential marketable product.

## B. Protein Adsorption on Elastomeric Surfaces

The following unpublished experimental work, performed by Kiang P (The West Co), Wang YJ (Scios Nova), and Kao P (Mayo Clinic), was presented by Dr. Kiang at the Parenteral Drug Association Course, Malvern, PA, 1995 (31). Some of their work is reproduced here with their permission (see Tables 2–5 and Figs. 1–5).

Some of the major protein instability mechanisms are reviewed and referenced in Chapter 3 of this book and some of the protein adsorption factors are found in Table 2.

The results from this brief study indicated the need for selective rubber closures coated with an inert, flexible polymer when formulating conventional and biopharmaceutical drug substances in order to achieve the longest shelf life. The benefits that could be derived from coated rubber stoppers may be several.

**TABLE 2** Protein Adsorption Factors

Ionic interactions
Hydrogen bonding
Molecular size and conformation
Competitive adsorption
Van der Waals forces
Surface roughness
Solvent effects
    ionic strength
    pH
    polarity
Charges on the surface
Surface tension
Crystallinity of surface
Tacticity
Crosslinking
Molecular weight
Concentration effects

*Source*: Kiang, Wang, Kao. PDA Short Course, Oct, 1995, Malvern, PA.

- Minimize the amount of adsorption onto the rubber stopper surface
- Minimal reactivity of the active drug substance with the coating on the rubber stopper
- Reduction of particulate generation during washing
- Reduction of extractables

**TABLE 3** Proteins and Peptides Used

|  | Molecular weight |
|---|---|
| Adrenocorticotropic hormone (ACTH) | 4,500 |
| Porcine insulin (PINS) | 6,000 |
| Human gastrin | 2,099 |
| Bovine parathyroid hormone (PTH) | 9,500 |
| Goat immunoglobulin G (IgG) | 155,000 |

*Source*: Kiang, Wang, Kao. PDA Short Course, Oct, 1995, Malvern, PA.

**TABLE 4**  Experimental Method

PBS buffer solution
   0.02M Sodium phosphate
   0.0145M NaCl
   0.02% NaN$_3$
   pH 7.2

PBS buffer used to dissolve
proteins or peptides

|  | Conc. | Specific activity |
|---|---|---|
| ACTH | 24 pg/mL | 440$\mu$Ci/$\mu$G |
| Insulin | 470 pg/mL | 140$\mu$Ci/$\mu$G |
| Gastrin | 47 pg/mL | 700$\mu$Ci/$\mu$G |
| PTH | 495 pg/mL | 200$\mu$Ci/$\mu$G |
| IgG | 17 pg/mL | 30$\mu$Ci/$\mu$G |

All proteins were labeled with $^{125}$I and detected using a Gamma
Counter

Test procedure
   10 of each stopper (10 cm$^2$/stopper) packed into polypropylene
   beakers. 30 mL of the tracer solution added to cover all stoppers.
   Incubate at room temperature for 48 hours, analyzing in duplicate
   500$\mu$L aliquots after 0, 1, 2, 4, 8, 24 and 48 hours.

*Source*: Kiang, Wang, Kao. PDA Short Course, Oct, 1995, Malvern, PA.

- Reduction of particulates during penetration with a hypodermic needle. Particles or coring effect with multidose vials
- The possibility of eliminating stacking effects during washing
- Potential elimination of silicone for stopper lubrication
- Last, but not least, improve machinability during manufacturing

**TABLE 5**  Rubber Elastomeric Closures

20 mm stoppers (10 cm$^2$ surface area)
   Butyl rubber
   Natural rubber
   Butyl rubber with PURCOAT
   Natural rubber with PURCOAT
   Butyl rubber with Tefzel fluorocarbon film

*Source*: Kiang, Wang, Kao. PDA Short Course, Oct, 1995, Malvern, PA.

Stability parameters applicable for clinical or marketable drug development are found in Chapter 5 of this book. It is highly advisable that the formulation scientists working closely with regulatory personnel address all the latest requirements for successful compliance for an Investigational New Drug (IND), Product License Application (PLA) or, if applicable, a New Drug Application (NDA).

Present stability studies should be designed taking into account the harmonization requirements of the United States, European community, and Japan, under the big umbrella of the International Conference of Harmonization (ICH).

## REFERENCES

1. Hopkins GH. J Pharm Sci 1965; 54:138.
2. Anschel J. Bull Parenter Drug Assoc 1977; 31:47.
3. Parenteral Drug Association. Elastomeric Closures: Evaluation of Significant Performance and Identity Characteristics. Tech Methods Bull 2 1981.
4. Heinisch KF. Dictionary of Rubber. New York: Wiley, 1966.
5. Avis EK. Pharmaceutical Dosage Forms. Parenteral Medication, Vol I, Second Edition. New York: Marcel Dekker, 1992.
6. Parenteral Drug Association. Elastomeric Closures: Evaluation of Significant Performance and Identity Characteristics. Tech Methods Bull 2 1981.
7. Williams JL. Proceedings of the 1st International Conference on Thermoplastic Elastomers Markets and Technology. Princeton, NJ: Scotland Business Research, 1988.
8. Stern HJ. Rubber: Natural and Synthetic. New York: Palmerton, 1967.
9. Am Soc Test Material. Annual Book of ASTM Standards, Sect 9, Vol 09.01–09.02. Philadelphia: ASTM, 1989.
10. Wood RT. Bull Parent Drug Assoc 1980; 34:286.
11. Wang WJ, Chien YW. Sterile Pharmaceutical Packaging: Compatibility and Stability. Parenter Drug Assoc Tech Rep 5, 1984.
12. Wing WT. J Pharm Pharmacol 1955; 7:648.
13. Wing WT. J Pharm Pharmacol 1956; 8:734.
14. Wing WT. J Pharm Pharmacol 1956; 8:734.
15. Lachman L, et al. J Pharm Sci 1962; 51:224.
16. Boyett JB, Avis KE. Bull Parenter Drug Assoc 1975; 29:1.
17. Milano EA, et al. J Parent Sci Technol 1982; 36:117.
18. Milano EA, et al. J Parent Sci Technol 1982; 36:232.
19. Mondimore D, Moore C. J Parent Sci Technol 1983; 37:79.
20. In: Smith DR, ed. The Blue Book. Akron, OH: Lippincott & Peto, 1984.
21. Plank AR. Rubber World. May, 1982; 186:35.
22. Code of the Federal Regulations, Title 21. Washington, DC: US Government Printing Office, April, 1988.

23. Keim FM. Elastomeric Closure Formulations: Composition, Development, Evaluation and Control. 19th National Meeting of the Academy of Pharmaceutical Sciences, Atlanta, GA, 1975.
24. Hopkins GH. Bull Parent Drug Assoc 1968; 22:48.
25. Generic Test Procedures for Elastomeric Closures. Inf Bull No 2, Philadelphia: PDA, 1979.
26. Hopkins GH. Bull Parent Drug Assoc 1969; 23:105.
27. Kapoor J, Murty R. Pharmaceut Technol Nov, 1977; 1:53, 80, 83.
28. Kay AI. Pharmaceut Technol May, 1977; 7:54.
29. Mattson LN, et al. J Parent Drug Assoc 1980; 34:436.
30. Murty R, Satyanarayana. J Pharm Tech 1993; 3:62.
31. Kiang P, Smith E. Parenteral Drug Association Course, Malvern, PA, 1995.
32. Knapp JZ, et al. J Parenter Sci Technol 1984; 38:128.
33. Introducing the N Series (prod broch). Phoenixville, PA: The West Co, not dated.
34. Omniflex, The Inert and Flexible Total Coating for Pharmaceutical Rubber Closure (tech broch). Pennsauken, NJ: Helvoet Pharma, 1993:2-09.

# Index

Printed in the United States
78870LV00002B/115-123

9 780367 4009